SCIENTIFIC AMERICAN

THE BIG IDEA

Editor DAVID H. LEVY is the author of several books, including *Eclipse: A Voyage to Darkness and Light*, *More Things in Heaven and Earth*, *The Man Who Sold the Milky Way*, *The Ultimate Universe*, the bestselling *Skywatching*, and *Sharing the Sky*. After Carl Sagan's death, Levy was chosen to become the new science columnist for *Parade Magazine*. As a respected astronomer, his most notable accomplishment is the co-discovery of the famed Shoemaker-Levy comet. He chronicled his account of the spectacular collision of the cometary fragments with the giant planet in *Impact Jupiter*. Levy lives in southern Arizona with his wife, Wendee.

D1510621

AVAILABLE NOW

COMING SOON

SCIENTIFIC AMERICAN
THE BIG IDEA

DAVID H. LEVY

Editor

PETER JEDICKE

Editorial Consultant

ibooks

new york
www.ibooks.com

DISTRIBUTED BY SIMON & SCHUSTER, INC

CONTENTS

Ideas that sparked some great debates, such as: How boulders were moved across the world, the nature of asteroids, the theory of spontaneous generation, the existence of humans beyond the biblical timeframe, the age of the world, the origin of life on Earth.

Theories and ideas that seemed right at the time, such as: The flammable nature of Earth's atmosphere, damming the Connecticut river, alien habitation of the Sun's atmosphere, turning the Arabian desert into an ocean, bringing metrics to America, harvesting ice from the Hudson River.

Bizarre ideas that had their day, such as: using human corpses to incubate chicken eggs, "electric men" built in Niagra Falls, a coin-operated public hairbrush, two-way vocal communication with fish, growing mustaches to prevent death by consumption, indoor bee hives.

INTRODUCTION

BY DAVID H. LEVY

It is amazing how much a three-year-old can teach you. Last Thursday our granddaughter Summer took us to the county fair. Summer, our daughter Nanette, my wife Wendee and I made quite a foursome as we hurried up the metal ramp to the Ferris wheel and handed our tickets to the attendant. We plunged into a bowl, shaped a bit like a coffee cup, and waited for the action to begin. We were tired on that hot afternoon: I had been up most of the night before taking photographs of the night sky in a search for new comets, and then most of the day working on a book. The pace of early 21st century life was going strongly for each of us. Nanette and her husband Mark had just moved to Albuquerque and were concerned about selling their old home and buying a new one.

Then, with a sudden jolt, the mighty Ferris wheel began to turn. As we ascended into the sky the Earth grew more distant. Nanette forgot about her home, Wendee and I forgot about our deadlines, and Summer just stared wide-eyed at the panorama of the landscape that was receding beneath her. It was a wonderful experience. We were on top of the world! Amazing how some steel, bolts, gears, chains, and a motor can cheer you up.

As the wheel continued its turns, my thoughts went back to a long gone event, reported in the July 1, 1893 issue of *Scientific American*. It reported the opening in Pittsburgh of the "wonderful machine" designed by engineer George W. G. Ferris. As hundreds of invited guests listened to speeches and congratulated the engineer, the wheel was ready for its first turn. And what a wheel it was! Carrying 36 cars, each sitting 40, the wheel could carry 1,440 people into the air at a time. "On the day the wheel was first started," the article went on, "5,000 guests were present at the inaugural ceremonies, all of whom were given a ride on the great wheel. The motion of the machinery is said to be imperceptible.

"The 36 carriages of the great wheel are hung on its periphery at equal intervals. Each car is 27 feet long, 13 feet wide, and 9 feet high. . . . To avoid accidents from panics and to prevent insane people from jumping out, the windows will be covered with an iron grating."

"Papo!" Summer pulled my arm and broke my reverie with the past. Her tiny arm pointed out the vast expanse of the desert below, right to the purple-mountained majesty of the Santa Ritas to our south. Down we went to kiss the desert once more, then up, up, and away once more. "Moon!" Summer shrieked as she caught a glimpse of our planet's satellite beyond the clouds.

Finally, our journey was over. We got out of our car and descended the ramp, finally back on terra firma. But somehow, our cares and deadlines had been left aboard that relic of technology from a century ago. The ride above the world had let us take on our own worlds with renewed energy.

BIG IDEAS

The pages that follow are a pageant of ideas that have tried to improve our knowledge, or make life more convenient or safer. All these ideas have appeared briefly in the pages of *Scientific American*. Many started out and ended as wonderful ideas, never going further than the pages of the magazine. Others, like George Ferris's wheel, made it into the history books along with the names of their inventors. These ideas were not always about Ferris wheels and mighty bridges—they include many things that make life easier, more fun, safer, or just plain more interesting.

Presented here is a universe of ideas and achievements by people with a sense of imagination, and of the bizarre: comet discoveries, thoughts on tree rings, fire extinguishers, laying transatlantic cable, discovering the Milky Way, microfilm, the telephone, the electric light the invention of aspirin, the nature of sunspots, slide projector, laws of heredity, noble gases, the escalator, the parachute, the greenhouse effect, the Polaroid camera, damming the Connecticut River, divining rods, the origin of life, installing coin-operated automatic hair brushes, and calculating the age of planet Earth. Just to name a few.

All ideas are not created equal. In between the ongoing great debates about how life began, and the age of our planet, dropping in a nickel to activate a hair brush probably was not an idea destined to rock the sands of time. But in a sense it was. Think of the person who came up with the idea. Was he rushing to an appointment, trying to catch a bus on a windy day, and suddenly realizing that he was a mess? While others would mope around, bemoaning the lack of something to help us freshen up, he did something about it. The result made it to the pages of *Scientific American* in June 1902.

OF SELENIUM AND COMETS

Most of us are aware of selenium, an element that occurs naturally in soils. We know that selenium is a mineral vital for good health, that we are exposed to it in our foods and can take it as supplements to help prevent certain cancers. As alluded to in the pages of *Scientific American*, its early history touched on its interesting electrical properties: "Mr. Willoughby Smith has been making a series of electrical experiments with selenium." *Scientific American's* March 1873 issue depicted the chemical element.: "Sticks of selenium were connected with platinum wire and hermetically sealed in glass tubes. The electrical resistance of some of these sticks was very great, the resistance of other sticks was much less. He was at a loss to account for this lack of constancy until after various trials he found it was due to the action of light. When the sticks of selenium were shut up in a box as to exclude light, the electrical resistance was highest and remained constant, but when the cover was withdrawn and light was allowed to fall on the sticks, the electrical resistance diminished according to the intensity of the light. These very singular observations may lead to new and useful discoveries."

Thanks partly to this article, chemists stepped up their experiments with selenium. Astronomers even experimented with the ability of the element to register the low light levels of stars seen through their telescopes, and thus marked the beginning of the use of selenium as a light gathering device in astronomy. What the editors of

Scientific American couldn't know was that eighteen years after that small announcement, selenium's interesting characteristic would be used as a practical joke. It was played against the famous comet discoverer E. E. Barnard. On March 8, 1891, while enjoying his morning coffee, Barnard saw this headline in the *San Francisco Examiner*'s human interest section:

ALMOST HUMAN INTELLECT
An Astronomical Machine That Discovers Comets All By Itself

The Meteor Gets in Range, Electricity

Does the Rest.

As Barnard read on, he learned to his amazement about his own fictitious invention of a telescope equipped with a selenium cell that would search the sky by itself. It described how comet seeking is a difficult and time consuming activity, and how wonderfully easy this new machine would make it. On paper at least, the procedure was simple. The device would examine the spectrum of everything: "Stars, nebulae and clusters innumerable crowd into the field [of view of the telescope] with every advance of the clock," the article went on, "but the telescope gives no sign of their presence . . ." But should any object showing the "three bright hydrocarbon bands" of comet light appear, the device as described by the paper would allow that light to pass through and hit a selenium band, closing a circuit and setting off an alarm in Barnard's bedroom. The sleepy astronomer would rush upstairs to the telescope, make a simple visual confirmation of the new comet, report it, and return to bed.

Barnard was livid. In letter after letter he tried to persuade the *Examiner*'s editors that he had never thought of such a device. Whoever perpetrated the hoax was careful enough to warn the newspaper that Barnard, a shy man, would doubtlessly deny everything, and the paper refused to print the scientist's disavowals. Meanwhile news of this marvelous use of selenium was spreading across the country. Although New York comet hunter Lewis Swift had his doubts that such a device would work, even he wrote to Barnard asking for all the details. "It takes my breath away," he gushed, "and makes my hair stand straight towards the zenith to think of it. Although the article appears somewhat fishy," he went on, "I am inclined to think it is still another of the marvelous inventions of the 19th century."

Almost two years passed before Barnard was able to assure the *Examiner* that it—and he—had been had. Finally convinced, the paper apologized on February 5, 1893: "The *Examiner* seizes the opportunity to express contrition for the annoyance which it caused this eminent scientist by printing some time ago an account of a highly ingenious, but non-existent, machine for scanning the skies and catching wandering comets on the photographic plate." At the end the paper bequeathed him "all the new moons and comets that may be necessary to his happiness." That same day the paper ran an article on "How to Find Comets"—a real one this time, by Barnard himself.

So Selenium, it turns out, was not the miracle cure for the hunter of comets, who stays up well into the night to search for comets visually or using photographs. But work on selenium went on. In May, 1903, the magazine announced that a Mr. Ernest Ruhmer ". . . has constructed an apparatus employing the selenium cell." "Some little discussion has taken place recently," wrote the editors of *Scientific American* three years later in November 1906, regarding the possibilities of " . . . 'seeing electrically.' Selenium when carefully prepared becomes an electrical conductor whose resistance is affected by incident light."

In November 1916, *Scientific American* prognosticated that in the future, a selenium cell with a 100-square centimeter surface could theoretically record the light of a 28th magnitude star—hundreds of times fainter than any star that could then be recorded by conventional photography—but the time exposure to accomplish this feat would have to last several nights. It is only in recent years that astronomers have been able to accomplish that feat. The power of computer chips was necessary, but to capture a star that faint a telescope had to be hurled into space. When looking through the stories and quips these pages offer, you may find other examples of how a farfetched dream finally became reality, but using very different routes from what originally looked so promising.

WILD IDEAS BY EMINENT SCIENTISTS

This book is marked by some ideas that seem strange, obscure, or just plain impossible. The perpetrators of these ideas are, for the most part, men and women with a little too much imagination, whose accomplishments are limited to their offbeat notions. But not all. The well-known astronomer John Herschel, whose meticulous observations of the southern sky brought him fame and a knighthood, made a suggestion that seems preposterous today. The Sun, he proposed, may be inhabited. From *Sci Am's* July, 1854 issue: "Between its luminous atmosphere and its surface there may be interposed a screen of clouds, whereby its inhabitants may no more suffer from intense heat than those who live in our tropical regions." The editors concurred, noting: "This may be so, as we all know how much the heat of the sun's rays, in the hottest days of summer, are modified by an interposing cloud or a swift passing breeze." No matter that no thought was given to how hot the Sun must be, 93 million miles away, to light up the entire solar system; if it was Herschel's idea, it had to have merit.

Actually, Herschel's notion came more than 60 years before another knighted English astronomer, Sir Arthur Eddington, would discover that our Sun, like other stars, gets its power by sustained nuclear fusion reactions that heat it up to millions of degrees.

LOOKING INTO THE FUTURE

Most of us followed closely the process of the presidential election of November 2000, the result of which was cast in doubt thanks to poor ballot design in Florida. This would really have surprised the editors of *Scientific American* from more than a century ago: In June 1895, the magazine announced that the "the days of ballot box stuffing and other modes of cheating at elections appear to be numbered." They announced the invention of a voting machine where names of Democratic Party candidates appear in yellow, republicans in red, and prohibition candidates in blue. "To the right of each name is a little knob which [the voter] must press in order to register his vote—the machine does the rest." Isn't it interesting that inventions are often seen as cure-alls? The technology of the voting machine couldn't be simpler, and yet 105 years later, they left us with a maze of hanging and dimpled chads and an election that took five weeks to settle.

So many other events, ideas, and suggestions have come our way—some amusing, some deeply prophetic, some just plain weird. Some, like the galvanic psychometer, were clearly on the right track; this early lie detector premiered in March 1909. Two years later, Percival Lowell's observations of the planet Mars confirmed "beyond a doubt" that the planet Mars has canals that are probably the result of construction by a civilization indigenous to the red planet. These canals, it turns out, did not bear scrutiny by the Mariner 4 spacecraft that visited Mars 50 years later to reveal many craters but no canals.

Back on Earth, the prefabricated home was introduced as the home of the future during the depression year of 1935. "You can now acquire a full-size, life-long home simply by ordering from a catalog and telling the dealer where to erect it. When you take possession a month later, the refrigerator will be making ice cubes, the heating plant will be throwing cool, conditioned air through the rooms and as likely as not, the radio will be playing. This is *the* prefabricated house." The premanufactured home is still with us, but asbestos fibers are not. In January 1947, asbestos piping was heralded as a material that would keep water supply and sewage pipes free from corrosion.

A PANORAMA OF IDEAS AND PROGRESS

Ideas move us forward, both physically and intellectually, but they evolve just like any life form. Sometimes that evolution is quite slow. When, for example, did the United States House of Representatives pass legislation making the metric system legal? When did *Scientific American* announce that "within a few days the action of the House will be confirmed by the Senate, when the metrical system will become the law of the land"? That bit of sensible news appeared in the magazine's May 1866 issue. Almost 140 years later, the metric system, at least for the measurement of distance, is clearly posted on signs on one 100-kilometer stretch of Interstate 19 between Tucson and Nogales. Evolution works, but it can be very slow.

When we read this book from the point of view of what might have been, we might come away disappointed. There are no intelligent, peaceful beings on Mars who overcame their differences to build a planetwide system of canals to save their civilization. Asbestos, for all its promise, was a dangerous material that had to be removed at great expense from every building in which it was installed. The premanufactured home, though still popular, has many disadvantages. The introduction of the English sparrow into the United States in 1862 was done with the best of intentions, but by 1882 was seen to have wreaked havoc on wheat fields, a problem we still have.

As we page through this tree of humanity's creativity, we certainly find that some of its branches are knotted and knurled, pointing downward or leading nowhere. But through those branches, others soar into the heavens. August 1858: the completion of the first trans-Atlantic telegraph cable. April 23, 1881: Verea's calculating machine can do all four arithmetical functions. October 17, 1885: the development of flexible film to replace photographic glass plates. March 1899: the Catherine Bruce telescope is set up near Arequipa, Peru, far away from city lights. March 1904: "After the motor device was completed," the magazine quotes a Mr. Wilbur Wright, "two flights were made by my brother and two by myself on December 17 last." (This invention would be followed by the showing in November 1921 of the first in-flight movie, and the introduction of automated flying in December 1932.) And finally, in May 1949, *Scientific American* reported the launching of a German V 2 rocket topped by a WAC Corporal second stage that soared 250 miles above the Earth.

I spoke with three people about that launch from so long ago. One was Clyde Tombaugh, whose discovery of Pluto was reported in the July 1930 issue. As he watched the rocket roar into the night, he visualized a world from which people could travel routinely into space, to the Moon, and even to the planets. Next to him stood his brother-in-law, James Edson, whose work also helped that rocket break the bonds of Earth. If we could depart our own world, he thought, could we not achieve anything? The third person, our granddaughter Summer, was born almost half a century after that launch, and her launch into space was powered not by a rocket but by Ferris's great wheel. What inventions, what discoveries, will Summer enjoy in her lifetime? What will her own contribution be? As our

Ferris wheel came down to Earth for the last time, we could almost be with James Edson and that distant launch in the early morning New Mexico sky, being with him as he composed in his mind the verses, reaching down to Summer's generation, about humanity's ultimate triumph:

Now, we've tracked a hundred mightier jets beyond the azure sky,
And a swarm of circling satellites as they went wheeling by,
And we were few, and we grow gray, and some of us are gone,
But our dream still lies beyond the skies,
And our hearts are bold like that bird of old,
That flew to meet the dawn.

SCIENTIFIC AMERICAN

THE BIG IDEA

CHAPTER ONE

FEBRUARY 1846

"A Mr. Philips of London has introduced an apparatus for the instantaneous extinguishment of fires. The principle of his fire annihilator is to project upon the fire a gaseous vapor which has a greater affinity for the oxygen of the atmosphere than the burning combustibles, and consequently extinguishes the fire by depriving it of the element oxygen, on which combustion particularly depends."

OCTOBER 1846

"Animal magnetism, with all its boasted advantages in rendering people insensible to pain, appears likely to be superseded by a discovery of Dr. William T. G. Morton, of Boston. It is no other than a gas or vapor, by the inhaling of a small quantity of which, the patient becomes immediately unconscious, and insensible to pain, thus giving an opportunity for the most difficult and otherwise painful surgical operations, without inconvenience."

DECEMBER 1846

"Urbain Leverrier's new planet [Neptune] is two hundred and thirty times as large as the earth, being the largest of the system. This discovery is perhaps the greatest triumph of science upon record. A young French astronomer sets himself at work to ascertain the cause of the aberrations of the planet Herschel [Uranus] in its orbit. He finds that another planet of a certain size placed at nearly twice the distance of Herschel from the sun would produce precisely the same effects he noted. He calculates its place in the heavens, with such precision, that astronomers, by directing the telescope to the point where its place for that evening is indicated, have all succeeded in finding it."

MARCH 1847

"Among all the new inventions and discoveries that are astonishing the world, we have heard of none which promises to be more useful and acceptable, at least to ladies, than 'The Essence of Coffee,' which is now offered to the lovers of that beverage. It is the genuine stuff, put up in bottles, at a low price. You have only to put a tea-spoon full into a cup of water containing the usual complement of sugar and milk, and you have a cup of superior coffee without further trouble."

APRIL 1847

"Philadelphians are in a high state of excitement, respecting the newly invented 'baby jumpers.' Imagine a cord fastened to the ceiling, and thence diverging into several cords, which are fastened to a child's frock by attachments to the belt. The cord is elastic, and the child may be left to itself and will find its own amusement in the constant jumping up and down and about, which its movements occasion."

MAY 1847

"A number of cabs with newly invented wheels have just been put on the road in London. Their novelty consists in the entire absence of springs. A hollow tube of India rubber about a foot in diameter, inflated with air, encircles each wheel in the manner of a tire, and with this simple but novel appendage the vehicle glides noiselessly along, affording the greatest possible amount of cab comfort to the passenger."

AUGUST 1847

"A gentleman in Baltimore has invented a Meat Safe, which promises to be most important. It consists of a chamber, so cut off from the influence of heat as to be at a degree or so above the freezing point. The ice, which is the preservative power, is replenished but once a year. The temperature is so low that the rotting as well as the over-ripening of fruits is prevented. Persons engaged in the bacon business can protect their meats from the inevitable effects of warm weather. The theory that cold was a preserver has long been maintained, but this invention has for the first time practically tested its correctness."

SEPTEMBER 1847

"On August 19, electrical cabs began to ply for hire in the streets of London in competition with the ordinary hackney carriages. The new vehicle resembles very closely a horseless and shaftless coupé, carried on four wooden solid rubber-tired wheels. A three-horsepower motor is supplied with current by 1,400 pounds of storage batteries. The cabs can travel up to thirty-five miles per charge and at speeds up to nine miles per hour. It is intended to have electric supply stations at other parts of London besides that at Juxon Street, Lambeth."

DECEMBER 1847

"The Pittsburg Gazette says: Messrs. Blackstock and Co. have made a trial of a smoke preventive apparatus, in their Cotton Factory in Allegany city. The experiment has proved successful. While the chimneys of the neighboring factories were vomiting forth clouds of black smoke that darkened the atmosphere on one of the finest Indian Summer days we have seen, the Smoke Preventive in the cotton factory consumed all the parts of smoke that dropped like rain from other points around us."

JANUARY 1848

"The eclipses of the moons of Jupiter had been carefully observed and a rule was obtained, which foretold the instants when the moons were to glide into the shadow of the planet and disappear, and then appear again. It was found that these appearances took place sixteen minutes and a half sooner when Jupiter was on the same side of the sun with the earth than when on the other side; that is, sooner by one diameter of the earth's orbit, proving that light takes eight minutes and a quarter to come to us from the sun."

AUGUST 1848

"The Boomerang is a curious instrument used as an offensive weapon by the blacks of Australia, and in their hands, it performs most wonderful and magic actions. The instrument itself is a thin curved piece of wood up to three feet in length and two inches broad—one side is slightly rounded, the other quite flat. An Australian black can throw this whimsical weapon so as to cause it to describe a complete circle in the air; the whole circumference of the circle is frequently not less than two hundred and fifty yards."

FEBRUARY 1850

"A French savant, M. Fiqueau, has just discovered a method of measuring the speed with which light travels, without any resort to the regions of astronomy. Two glasses are fixed opposite each other, so that the focus of the one (having a mirror) reflects a ray of light, starting from the focus of the other back to that focus again. A disc is provided to revolve at this point; and the eye, observing whether the ray appears, or is eclipsed, knows whether it has encountered a tooth of the disc, or one of the vacant spaces between the teeth; and thus elements are found for a calculation which shows the speed of light to be very nearly the same as that arrived at by the astronomical calculation of Bradley or Roemer."

APRIL 1850

"Dr. Alexandre, from Paris, has lately brought out a sub-marine boat, in which a company of persons can go down to the bottom, have communication with the land, performing any sort of work by digging or otherwise, and return to the surface at will."

MAY 1850

"Chloroform has been employed in Edinburgh in from 80,000 to 100,000 cases, without a single accident or bad effect of any kind traceable to its use. It saves many lives which otherwise would sink under the nervous shock which is experienced from a severe operation undergone in a state of consciousness. At the same time, chloroform has received the sanction and recommendation of the most authoritative bodies in France and the United States. Nevertheless, the public of London is almost wholly denied the vast benefits of this agent, purely through the prejudices of profession."

JULY 1850

"The binary stars: To Sir William Herschel the honor of discovering this extraordinary combination of the heavenly bodies is due. That great man remarked that there were many instances of two stars being placed so close together as to appear to the eye as one, it being only by means of the telescope that their separate orbs could be descried. Extended observation soon showed that this combination occurred far too frequently to be the mere effect of accidental similarity of direction; there is no position in astronomy better established than the fact that two, three, or more stars may be found in combination revolving round each other and exercising a combined influence on the planetary systems relating to each."

NOVEMBER 1850

"The existence of a third ring around Saturn, which had been for some time suspected, was ascertained by the astronomers at Cambridge. It is interior to the two others, and therefore its distance from the body of Saturn must be small."

DECEMBER 1850

"At a sitting of the French Academy of Sciences M. Claude Bernard submitted a communication on the functions of the liver in man and in animals. 'I am about,' he said, 'to demonstrate experimentally that the presence of sugar in animal organisms is a constant and indispensable fact in nutrition, caused in the liver by a special function of that organ. The liver has thus two functions: to wit, on one hand, the secretion of bile and on the other, the production of sugar. This latter function begins to be performed before birth—for I have ascertained the presence of sugary matter in the liver of the fetus of mammals and of birds at different periods of fetal life. The sugar from the liver has all the characteristics of glucose.'"

AUGUST 1851

"A great reaping match was held on July 24 in Essexshire, and thither were invited all the reaping machines exposed at the Great Fair. A number were tried but proved abortive in their attempts to work well. It was then the stout but unprepossessing machine of Mr. McCormick made its appearance. Those who estimated the worth of the machines by a polished piece of brass here, and a burnished piece of steel there, shook their heads as the driver mounted his seat; but with a snap of his whip he started his team, applied his hand to the lever of his clutch, and away he went, sweeping a wide swath and raking it up on the platform at one operation, with such a velocity as to elicit repeated cheers from the onlookers. The success of this experiment will lead to the introduction of the American power reaper into Britain."

OCTOBER 1851

"A new style of vessels named clippers have come into existence for sea voyaging within the past two years. They are built more for making fast passages than for carrying cargo. They are beautiful in shape, and carry a great amount of sail. The vessel of this class which created the greatest excitement in this city was the *Flying Cloud*, built by Donald McKay, of Boston, and she has made the fastest run to San Francisco. She made the voyage from New York in 80 days. In one day she runs 374 miles, averaging about 16 knots per hour. This speed beats our fastest Atlantic steamers."

NOVEMBER 1851

"A very interesting experiment has been lately performed at the Hotel Dieu of Lyons. A female was brought into the hospital who had been seized with violent hemorrhage. Her condition seemed desperate. Death appeared imminent, inevitable. Doctor Delorme suggested transfusion. This was at first combatted by the other physicians as offering no chance of success, but was finally assented to, as, the case being a desperate one, it could do no harm, even if it did no good. The proper vein in the arm of the sufferer was then opened, and a fine canula, or tube, was introduced to some length. The other end of the tube was then fitted to the syringe in which was the necessary quantity of pure human blood. The operator then gently forced into the veins of the dying woman the revivifying fluid. Soon after, she recovered, in a great degree, her senses and eyesight. The patient has since been regularly improving, and the cure may be set down as complete."

JANUARY 1853

"A correspondent states that the operators of the telegraph running between Buffalo and Milwaukee, working under Morse's patent, have for some time past discontinued the practice of recording the signs, and have instead thereof received, their messages by sound. The operator sits by the table in any part of the room where the message is received, and writes it down as the sounds are produced. The different sounds are made by the striking of the pen lever upon a piece of brass; thus three raps in rapid succession are made for the letter 'A,' two raps, an interval, and then two raps more are made for 'B,' and so forth."

APRIL 1853

"Mr. Charles Goodyear has recently taken out a patent in England for a new compound composed of india rubber and coal tar vulcanized with sulphur. Coal tar is heated in an open boiler until it acquires the consistency of melted rosin, when it is mixed with india rubber in proportions which may vary according to the character of the material to be produced for a specific purpose. It is mixed with sulphur and then heated to vulcanize it."

JANUARY 1854

"If all the reports which have come to us recently from abroad with respect to new discoveries in making candles are true, all our whaling ships will soon be laid up in port, or converted into *coal grunters.* Candles are now manufactured in Scotland from coal. There is a quarry about 12 miles to the west of Edinburgh which rests immediately above a thick bed of dark-colored shale. The shale has been found to be exceedingly rich in an inflammable substance, resolvable into gas and tar, and which, from the paucity of its chemical affinities, has received the name of parafine. Of this substance beautiful candles are made, in no degree inferior to those of wax."

OCTOBER, 1854

"By some recent experiments of Regnault, in Paris, the old hypothesis of heat being a fluid seems to be settled in the negative, and the phenomenon of heat, like sound, is attributed to a vibratory motion in bodies. In a recent lecture, he stated that if hot air in a vessel like a glass globe be allowed to expand into another empty vessel kept in a water bath at the same temperature, there would be neither an elevation nor depression of the temperature of the air, although it were allowed to expand to 10 times its former bulk. But if that air be allowed to escape to do work such as to move a turbine, or pump, the cooling increases according to the work done. 'Consequently we find,' he says, 'that the useful work done is more nearly expressed by the heat lost in the fall of temperature, in proportion as the machine is perfect.' "

SEPTEMBER 1856

"At the meeting of the British Association for the Advancement of Science, held in Cheltenham, Eng., last month, Henry Bessemer, of London, read a paper on a new method of making malleable iron from pig iron, which deserves the attention of our iron manufacturers, as the process is very original, is stated to be perfectly successful, and destined to revolutionize the processes of manufacturing malleable iron and steel. The idea occurred to him that if molten pig iron at a glowing heat were run into a chamber and a blast driven through it, the 5 per cent of carbon in it would unite with the oxygen of the blast, producing intense combustion, because carbon cannot exist at a white heat in contact with oxygen. He then put up a cylindrical vessel three feet in diameter and five feet high, like an ordinary cupola furnace, the interior of which he lined with fire brick. At about two inches from the bottom are inserted five tuyère pipes, having nozzles of fire clay. A blast of air of a pressure of eight pounds to the square inch is let into this cylinder a few minutes before the crude iron is allowed to flow into it from the blast furnace. The molten crude iron is then let in by its tap, and it soon begins to boil and toss about with great violence. Flames and bright sparks then begin to issue from the vessel's top; the oxygen of the air from the blower combines with the carbon in the metal, evolving a most intense heat producing carbonic acid gas, which escapes. By this simple process the heat generated is stated to be so intense that all slag is thrown out in large foaming masses, and all the sulphur is driven off, together with deteriorating earthy bases, so that the metal is completely refined—more pure than any puddled iron. It is also stated that one workman can convert five tons of crude pig into malleable iron in about 30 minutes by this process. Its advantages are painted in such dazzling colors that we are afraid to rely upon them implicitly. If they are such as Mr. Bessemer has described, a new era in the manufacture has dawned upon the world, and malleable iron will soon be reduced to a price but little above common pig."

OCTOBER 1856

"It is a remarkable fact in the history of the useful arts that asphalt, which was once so generally employed as a durable cement, should have almost fallen into disuse for thousands of years. Some attempts have been made in this city to make a concrete pavement of it, but for this purpose it is evidently not equal to stone flags, because it has had to be relaid, and now huge cracks are again seen in different parts of it. On the other hand some beautiful mosaic asphalt pavement has been laid down in the streets of Paris, and is said to be perfectly successful."

DECEMBER 1856

"A most remarkable new invention, which we have recently examined, is a small and neat hand-machine for printing. Its object is to print letter after letter, as a substitute for writing with pen and ink. The devices combined to execute the printing continuously in lines are ingenious. The letters of the alphabet, numbers, punctuation marks and spaces are so arranged that when a lever is pressed down, the letter is forced upwards and impressed on a sheet of paper. The paper is fed into the machine on a roller. When one line is printed, the roller is turned forward one notch, then pushed back to its starting position, and the machine is ready to print another continuous line. A band of paper of any length may be used in the machine, and the printed portion can be read as it is fed out."

JANUARY 1857

"The Rev. Dr. Livingstone has recently returned to England from an African adventure of the most dangerous and thrilling character. He has traced by himself the course of the great river Zambesi, in eastern Africa, extending 2,000 miles. This immense stream, whose discovery is the great fruit of the journey, is an enigma without parallel, for only a small portion of its waters ever reach the sea. Like the Abyssinian Nile, it falls through a basaltic cleft, near the middle of its course, which reduces its breadth from 1,000 to only 20 yards. Above these falls it spreads out periodically into a great sea filling hundreds of lateral channels. During his unprecedented march, alone among savages to whom a white face was a miracle, Dr. Livingstone was compelled to struggle through indescribable hardships. The hostility of the natives he conquered by his intimate knowledge of their character and the Bechuana tongue, to which theirs is related. He waded rivers and slept in the sponge and ooze of marshes, being often so drenched as to be compelled to turn his armpit into a watch pocket. He has brought back memoranda of the latitudes and longitudes of a multitude of cities, towns, rivers and mountains, which will go far to fill up the unknown regions in our atlases. Toward the interior he found the country more fertile and populous. Lions were numerous, being worshiped by many tribes as receptacles for the departed souls of their chiefs. The natives also worshiped idols, believed in transmigrated existence after death, and performed religious ceremonies in groves and woods. They were less ferocious and suspicious than the seaboard tribes, had a tradition of the Deluge, and had more settled governments. Some of them practiced inoculation and used quinine, and all were eager for trade. He has described a Quaker-like tribe, on the river Zanga, who never fight and never have consumption, scrofula, hydrophobia, cholera, smallpox or measles. Dr. Livingstone is nearly forty years old. His face is furrowed by hardship and thirsty fevers, and black with exposure to a burning sun. His left arm is crushed and rendered nearly helpless from the attack of a lion. His discoveries, in their character and commercial value, have been declared superior to any since the discovery of the Cape of Good Hope by Vasco da Gama. But greater than any commercial value is the lesson which they teach—that all obstacles yield to a resolute man."

MAY 1857

"For a long period astronomers unsuccessfully endeavored to determine the distance between the stars and the earth, and it is only within a comparatively short time that the interesting problem can be said to have been solved. The distance which separates us from the nearest stars is, according to M. Arago, about 206,000 times the distance of the sun from the earth—more than 206,000 times 95,000,000 of miles. Alpha, in the constellation of Centaur, is the star nearest to the earth; its light takes more than three years to reach us, so that, were the star annihilated, we should still see it for three years after its destruction. If the sun were transported to the place of this, the nearest star, the vast circular disk, which in the morning rises majestically above the horizon, and in the evening occupies a considerable time in descending entirely below the same line, would have dimensions almost imperceptible, even with the aid of the most powerful telescopes, and its brilliancy would range among the stars of the third magnitude only."

AUGUST 1857

"According to the observations made by M. Rudolphe Wolf, Director of the Observatory at Berne, it appears that the number of spots on the sun have their maximum and minimum at the same time as the variations in the needle of the compass. It follows from this that the cause of these two changes on the sun and on the earth must be the same, and consequently, from this discovery, it will be possible to solve several important problems in connection with these well-known phenomena, the solution of which has hitherto never been attempted."

AUGUST 1858

"The Atlantic Cable is laid! All hail to Anglo-Saxon genius! And two nations' heartfelt thanks to the noble—aye, and mighty—men of science, capital and energy, whose untiring zeal and indomitable perseverance have linked the hemispheres with the electric cord! But a month ago we announced the third failure in this enterprise, but we did not groan and lament. Cyrus W. Field, Professor Morse and all connected with the enterprise are great pacificators, great civilizers. The telegraph fleet met at mid-ocean on Wednesday, July 28th, and made the splice at 1 p.m. on Thursday, the 29th, and then separated— the *Agamemnon* and *Valorous* bound for Valentia, Ireland, and the *Niagara* and *Gorgon* for Trinity Bay, Newfoundland, where they arrived on August 4th. It is 1,698 nautical miles between the two telegraph houses, and for more than two thirds of this distance the water is over two miles in depth. On August 16th the first telegram, directed from Queen Victoria to President Buchanan, was received. On the evening of the 17th, New York and many other cities were brilliantly illuminated, fireworks were let off, and the people generally had a good time of it throughout the country; and here, to celebrate the event properly, the cupola and upper story of our City Hall were burned."

DECEMBER 1859

"Aluminum, the metal of which we have heard so much during the past few years, does not appear to be about to realize the brilliant expectations which were formed at the time of its first successful production in quantity, by St. Clair Deville. In its pure state it approaches somewhat to the color of silver; but it is found nearly impossible to free it from the certain foreign matters which become alloyed with it during the process of production. Aluminum is now sold in London for about $15 an ounce, being nearly as costly as gold, but, in consequence of its small specific gravity, an immensely larger bulk is given in an ounce of aluminum than in an ounce of gold. When aluminum becomes cheap, its nontarnishing property will bring it at once into varied and extensive use for many articles of domestic economy—tea utensils, spoons, knives and forks, door-knobs, etc. Verily, a millennium of housekeeping may not be far distant, for M. Deville has prophesied that aluminum will ultimately be cheaper than silver."

NOVEMBER 1860

"The railroad companies in France are about to put in operation a plan which cannot fail of being received with favor by the public. It is proposed to run, each week, a train of cars between distant points, for which tickets can be obtained in advance, and to which the companies will guarantee to admit only a limited number of passengers. All the places being occupied, the engine not carrying any 'dead weight,' the traveler can be transported at the price of merchandise. The companies, not only without loss, but even with a certain and calculable profit, will apply to these trains a tariff, the great cheapness of which cannot fail of producing an immense business. By this arrangement the fare is about one-fifth the usual price."

DECEMBER 1860

"It has been ascertained that our sun is one of an innumerable multitude of stars which are grouped together in one collection or system, separated from other stars in the universe. The general form of this stellar system is an irregular wheel, with a deep notch in one side, and with a portion of another wheel branching out from it. Our sun is situated pretty near the middle of the system. The dimensions are so vast that the plan has been adopted of stating the time which a ray of light would require to traverse them. In applying this measuring rod, it is found that, through the thickness of the wheel the distance is such that light would occupy 1,000 years, and through the diameter not less than 10,000 years, in making its passage! In some directions, indeed, the system stretches away into depths of space beyond the reach of the most powerful telescopes. If we pass through these inconceivable distances out beyond the boundaries of our stellar system, we find a region of empty space. Traversing this void space through distances which appall the mind by their immensity, we find other systems of stars probably similar to our own. And astronomers are now considering the possible relation of these several clusters to each other—whether there is not a system of systems! This is the most sublime problem which has ever engaged the attention of the human mind."

MAY 1861

"The most remarkable scientific event of modern times is the publication of a treatise on chemistry proceeding on the same plan in organic chemistry as has been adopted for a century past in mineral chemistry; that is, forming organic substances synthetically by combining their elements with the aid of chemical forces only. The author has been occupied with organic synthesis since he first devoted himself to chemistry. He is convinced that 'we may undertake to form, *de novo*, all the substances that have been developed from the origin of things, and to form them under the same conditions, by virtue of the same laws and by means of the same forces that nature employs for their formation.'"

JUNE 1861

"The Paris correspondent of the London *Photographic News* says upon the subject of zinc and steam: 'The employment of electricity as a motive power depends on its relative economy with steam, or the difference between the cost of zinc and coal: for in the electric battery it is the zinc that is consumed. But a remarkable feature in the question is that, while ordinary steam engines render only .052 of chemical power, the electromotive machine yields .20 to .25, which is enormous, and gives it an undoubted superiority over steam. Yet, even at this rate, electromotive power is 20 times dearer than that of steam. The question to be solved, therefore, is the economic production of electricity.'"

SEPTEMBER 1861

"When repeated charges of electricity are passed through a jar filled with atmospheric air or with pure oxygen gas, the oxygen acquires new properties. It emits a peculiar odor, it possesses extraordinary bleaching powers, and it has its affinities, or power of combining with other substances, very largely increased. Schönbein, who first discovered this fact, supposed that he had found a new substance, and he gave it the name of ozone, from the Greek *ozo*, odor; its most striking peculiarity being the odor that it emitted. We find in *La Répertoire de Chimie Appliquée* an account of some recent investigations that have revived the first idea of Schönbein, that ozone is not oxygen but a separate element. Messrs. Andrews and Tait, after a long series of observations, regard it as probable that oxygen is a compound substance, and ozone is one of its elements."

JANUARY 1862

"Oregon has no magnetic telegraph as yet, but it is arranged that before the middle of 1862 Portland shall be in communication with the wires of California, and through them with Chicago, New York and Boston. Sitka, in the Russian possessions, is only 900 miles from Portland; and when a line is completed between the two places, to connect with the Russian line, 3,500 miles long, soon to be undertaken between the Amoor River and Sitka, the circuit of the world will be complete."

MAY 1862

"A subterranean railway is now in an advanced stage of construction, running about four and a half miles under the city of London. It commences at Victoria Street, whence it passes eastwardly, having a large number of intermediate stations. On the occasion of a recent trip made through a portion of its length, the air was found to be perfectly sweet and free from all unpleasantness."

JUNE 1862

"Gen. McClellan's valuable adjunct to his corps d'armée—the Lowe reconnoitering balloon—is getting to be quite an institution. During a fight lately between the rebels and a force of Union troops, in which the latter were engaged in dislodging some batteries that had been erected, the balloon did effective service in directing the movements of our artillery. A telegraph wire, attached to an instrument on board, conveyed intelligence to our men what to do and what not to do, and corrected any mistakes made by the transmission of such messages, as 'too short,' 'just a little over,' 'fire lower,' etc. The enemy could not be seen by the men at the batteries, and our batteries in turn were hidden from the view of the enemy, the majority of whose shots fell wide of the mark."

SEPTEMBER 1862

"The *Great Eastern* arrived at her destination near Harlem in Long Island Sound with about 1,400 passengers and a general cargo. When passing Montauk Point she struck a sharp sunken rock, which opened a leak through which the water entered so fast that the pumps were unable to keep it down. Since the ship is divided into several watertight compartments by bulk heads, only one has been filled by the leak. Her bottom has been examined and will be repaired before she proceeds on her return voyage. The damage is but slight and none of the goods were injured."

MAY 1863

"Sir H. Davy, in his important and interesting experiments, found that light carbureted hydrogen, the most powerfully explosive of the gases, required about seven times its bulk of atmospheric air to be mixed with it to produce the greatest explosive effect. Practically, it can be calculated that from eight to nine times its bulk of air will produce the most explosive mixture of coal-gas; but the air and gas must be mixed previous to inflammation. No matter how rapidly the air may be supplied when the gas is burning, it will merely increase the fierceness of combustion; there will be no explosion."

JUNE 1863

"The magnetic needle is like a wind vane, as it serves to render visible the direction and intensity of that mysterious force which operates through the earth. Observations upon the magnetic needle reveal the fact that it will sometimes start and oscillate with great activity without any apparent cause, and it has been noticed that magnets in various parts of the world are always thus agitated at the same moment. It is believed that these phenomena have a connection with movements in the sun. Prof. Schuabe of Dessau has been watching the disc of the sun for nearly 40 years, and he has recorded the groups of spots which have appeared upon it. He has found that these occur in greater number in periods of about 10 years. They were noticed in 1848 and 1859, and in these two years great disturbances of the magnetic needle were observed. Magnetic storms are always accompanied by auroræ and earth magnetic currents. It appears that magnetic disturbances occur in the sun, in the earth's atmosphere and in the earth itself, at the same time and at regular periods. The mysterious force, 'magnetism,' seems to pervade the entire solar system, and perhaps the whole universe."

OCTOBER 1863

"The changes which have recently taken place in the use of fluids for artificial light have been rapid and astounding. Only a few years ago whale and lard oils were the common agents for this purpose; then these were superseded in great measure by that dangerous compound of alcohol and turpentine called 'burning fluid,' and again this agent was displaced by oil called 'kerosene,' distilled from cannel coal. Now this oil too has been superseded by petroleum—the natural product of wells situated in the valley of the Allegheny in Pennsylvania. How this fluid is produced in nature's laboratory is still a subject of speculation; in most respects it is similar to the oil obtained from coal, but it has been supplied so profusely and at such low prices as to have completely annihilated the manufacture of kerosene. In the course of two short years the petroleum trade has attained to gigantic proportions. In 1861 only a few hundred thousand gallons of it were exported; in 1862 about five millions of gallons; while during the past seven months of this year, ending in September, 21 millions of gallons had been exported."

FEBRUARY 1864

"Professor Agassiz lately delivered a course of three lectures before the Smithsonian Institution, and the greater part of the last one was devoted to a description of the phenomena which indicate that the continent of North America had at one time been overlaid by dense and unbroken masses of ice, moving from the North to the South. After stating the grounds on which the 'earthquake theory' was inadequate to explain the peculiar drift deposited on the surface of the continent from the Arctic to the 36th or 40th parallel of latitude, Prof. Agassiz estimated that the ice which deposited this drift and produced its other attendant phenomena must have been five or six thousand feet thick. But whence came the cold which produced such a thickness of ice? This query was answered by supposing that there had been injected into the sea from the subterranean fires of the earth below it a vast mass of melted material, thus generating an immense volume of vapor, which, escaping for ages into the upper air, was condensed and fell in the shape of snow and hail. By this mass of snow and hail the temperature of the earth's climate was reduced from the comparative warmth which preceded it, even in Arctic regions, and the world entered on the 'cold period,' which it was the object of the lecturer to describe and to account for while describing."

JANUARY 1865

"A lucifer match is now in the market that differs from anything hitherto in existence. Upon the side of each box is a chemically prepared piece of friction-paper. When struck upon this, the match instantly ignites; when struck upon anything else whatever, it obstinately refuses to flame. You may lay it upon a red-hot stove, and the wood of the match will calcine before the end of it ignites. Friction upon anything else but this prepared pasteboard has no effect upon it. The invention is an English one, and by a special act of Parliament the use of any other matches but these is not permitted in any public buildings."

FEBRUARY 1865

"The Paris correspondent of the *Chemical News* states that a curious experiment has been made by Dr. Reichenbach of Vienna. He believes in the existence of a cosmical powder or dust, which exists all through space and which sometimes becomes agglomerated so as to form large and small meteorites, whereas at other times it reaches the surface of our earth in the form of impalpable powder. We know that meteorites are mainly composed of nickel, cobalt, iron, phosphorus, etc. Dr. Reichenbach went to the top of a mountain that had never been touched by spade or pickax and collected there some dust, which he analyzed, and found it to contain nickel, cobalt, phosphorus and magnesia."

MARCH 1865

"M. H. Tresca has communicated a paper on the flow of solids under pressure to the French Academy, in which he details experiments to show that 'solid bodies can, without change of condition, flow after the manner of liquids, if sufficient pressure is exerted on them.' Mr. Tresca thinks that operations of this kind may explain cases of intrusion of one rock into another."

SEPTEMBER 1865

"Glycerine, as we all know, is the sweet principle of oil and is extensively used for purposes of the toilet, but it has now received an application of rather unexpected nature. In 1847 a pupil of M. Pelouze's, M. Sobrero, discovered that glycerine, when treated with nitric acid, was converted into a highly explosive substance, which he called nitro-glycerine. This liquid seems to have been almost forgotten by chemists, and it is only now that Mr. Nobel, a Swedish engineer, has succeeded in applying it to a very important branch of his art, viz., blasting. From a paper addressed by him to the Academy of Sciences we learn that the chief advantage which this substance, composed of one part of glycerine and three of nitric acid, possesses is that it requires a much smaller hole or chamber than gunpowder does, the strength of the latter being scarcely one-tenth of the former. Hence the miner's work, which according to the hardness of the rock represents from five to 20 times the price of gunpowder used, is so short that the cost of blasting is often reduced by 50 per cent."

DECEMBER 1865

"Some interesting facts respecting yeast have been brought before the Academy of Sciences by M. Bechamp in a note 'On the Physiological Exhaustion and Vitality of Beer Yeast.' The author washed and washed globules of yeast until they appeared to be mere envelopes of cellules and found that they still retain the power of changing cane sugar into glucose and setting up the alcoholic fermentation, which proves, he considers, that the property of setting up fermentation resides in the properties of the living cellule and is a consequence of the act of nutrition of this cellule."

DECEMBER 1866

"The Agricultural Committee of Sologne, France, has awarded the gold medal offered some time since to the inventor of a process which should enable French wines to be conveyed by land and sea, and preserved in any climate, without alteration in flavor. M. Pasteur, who receives the award, has succeeded in establishing the fact that the heating of ordinary wine to the extent of 50 degrees centigrade is sufficient to kill all microscopic vegetation, or the ferments by which it is produced, without affecting color or flavor, and to ensure the preservation of the wine in closed vessels for an indefinite period. The various morbid changes in wines are found to be due to various stages or phases of microscopic vegetation, which M. Pasteur has accurately described."

FEBRUARY 1867

"Mr. James Parker describes in Engineering an apparatus for propelling vessels by steam without an engine. The steam is issued in extremely small jets, each shooting into the center of an open pipe a quarter of an inch in diameter, conducting into a hot water chamber, into which the jet carries with it a current of compressed air. This compressed and heated air is admitted upon the surface of the water in closed tanks by the ordinary slide valve, and its force is employed to eject the water through propelling pipes."

MARCH 1867

"A volcano in the moon is said to be in active eruption. The crater called Linné has been lately observed to be obscured, and it is said that the same darkness was observed on this spot in 1788."

NOVEMBER 1867

"Artificial rubies, not mere copies in glass but made veritably out of the same substance—alumina—of which the natural gems are composed, have been produced by M. Ebelsman of the Sèvres Porcelain

Works near Paris. The process consists in employing a solvent, which dissolves the mineral or its constituents and may thus, either upon its renewal or by a diminution of its solvent powers, permit the mineral to aggregate in a crystalline state. Certain proportions of alumina, magnesia, oxide of chromium, or oxide of iron, and fused boracic acid are placed in a crucible made of refractory alumina enclosed in a second one, the whole being exposed to a high heat. The materials are first dissolved in the boracic acid; then as the heat continues the latter evaporates, the alumina and coloring matter combine, crystallize and present the exact appearance of the spinel ruby."

APRIL 1868

"Prof. Nobel of Hamburg, not entirely content with his former discovery, nitro-glycerin, has brought out another explosive, to which he has given the name 'dynamite.' Instead of being an oily liquid, liable to leak from the vessel in which it is confined and produce a spontaneously inflammable mixture with rags, shavings and other packing material, this powder resembles snuff in appearance. In a loose, non-compressed condition it does not explode but burns slowly with little smoke, an invaluable property in working closed mines or tunnels. A detonating cap is required to explode it. Recent California papers contain accounts of the prodigious power of this powder, demonstrated in some experiments tried in that state. They recommend it highly as being vastly more explosive and requiring much less drilling or preparation of the rock than gunpowder."

FEBRUARY 1869

"The practice of the transfusion of blood seems to be coming more prominently into notice at the present time than it has been for some years past. The *Medical Record* gives an account of a successful operation for the transfusion of blood recently performed by Dr. Enrico Albanese at the hospital of Palermo, Sicily. A youth aged 17, named Giuseppe Ginazzo, of Cinisi, was received at that establishment on the 29th of September last with an extensive ulceration of the leg, which in the end rendered amputation necessary, the patient being very much emaciated and laboring under fever. The operation reduced him to a worse state than ever, and it became apparent that he was fast sinking, the pulse being imperceptible, the eyes dull and the body cold. In this emergency Dr. Albanese had recourse to the transfusion of blood as the only remedy that had not yet been tried. Two assistants of the hospital offered to have their veins opened for the purpose, and thus at two different intervals 220 grammes of blood were introduced into the patient's system. After the first time he recovered the faculty of speech and stated that before he could neither see nor hear but felt as if he were flying in the air. He is now in a fair state of recovery."

APRIL 1869

"Bauxite is the name of a new mineral that has recently been discovered in France. It is reported to be a hydrated oxide of alumina, in which iron has been replaced by alumina. Bauxite is therefore a source of aluminum."

AUGUST 7, 1869

Caissons for Pier-Building

The illustration which we give this week shows a more economical method of building piers in beds of rivers, or under water. It shows a caisson or diving-bell, designed by Messrs. Burmeister and Wain, and adopted by them in building the piers of the new bridge in Copenhagen. The principal economy consists in having the caisson, or cylinder, of less cubic capacity than the finished pile of the piers, and in being able to take it away as each pile was built. When the excavation was made deep enough for a firm foundation, the building of the pile was commenced, and as it increased in height the caisson was lifted accordingly until the pile was about water-line, when the caisson was removed to the required position of the next pile, and so on, until the two piers, each formed of two piles, were completed. This plan of lifting the caisson avoided leaving the whole of the piles of the piers encased in ironwork, as in the piers of Rochester and many other bridges. This caisson was made of wrought-iron, 18-feet diameter at the lower part by 8 feet high, and above this to the air-chamber out of the water it was only 10 feet diameter. Just above the 18-feet diameter chamber there were two annular rings, or chambers—one to contain iron ballast and the lower one water ballast so that in sinking the caisson the water chamber was filled with water for weight in addition to the iron ballast in the annular chamber above. When they had excavated to the solid strata, a bed of concrete 3 to 4 feet thick was formed, and on this the remainder of the pile was built with granite facing filled in with brickwork. As the building of the pile proceeded, the caisson was lifted by means of the suspension chains connected with staging overhead, and by pumping air into the annular air-chamber to displace the water. The finished piles are about 18 feet diameter at their bases, and 16 feet diameter at their tops, by 30 feet high. The whole of the work below water line was done in the 18 feet by 8 feet chamber at the bottom of the caisson. Between the time of lowering it to the bed of the river and the completion of the first pile to water line was only twenty-eight days, and then the apparatus was moved into position for the next pile. In lowering it for the second pile, it unfortunately got upset, and caused so much delay that it took thirty-six days to complete this pile. The third pile was, however, finished in sixteen days, and the fourth in seventeen days.

JANUARY 1870

"During the year 1869 the extraordinary spectacle has presented itself of astronomers turning chemists. The telescope has been exchanged for the spectroscope, and mathematics has been laid on the shelf. Our observatories have become chemical laboratories. The physical conditions of the sun's atmosphere, its protuberances, and the source of its light and heat have been chiefly studied. All of the heavenly bodies have been brought down to earth for accurate analysis. No sooner was the computation of the time of the last total eclipse completed than the slate was thrown aside and preparations were made to photograph every stage of the obscuration and to study the light of the sun by means of the spectroscope. The chief preparations were chemical ones, and the most novel discoveries were in this department of science. We find hydrogen in the sun, and curious bands in the fixed stars and planets."

MARCH 1870

"The notion of sudden catastrophes at remote periods, by which mountain chains were upheaved or seas opened, is now generally abandoned. Sir Charles Lyell has succeeded in making the doctrine generally accepted that geological facts are to be explained by forces now at work; that the same power which now raises the coast of Scandinavia at the rate of a few inches in a century and depresses that of parts of New Jersey about as fast if it has time enough to work in will suffice to make continents of all the oceans and to submerge every continent; that the earthquakes and eruptions which have built up some mountains and islands in our own time need nothing but more time to build innumerable others."

CHAPTER TWO

APRIL 1871

"Not long since, the cable between Lisbon and Gibraltar was disabled. After considerable labor it was grappled in 500 fathoms of water. It had been supposed that at that depth the ocean is generally at rest and that there are no currents below 200 or 300 feet from the surface. When brought to the deck of the repair ship, however, there appeared on the cable most evident indications of chafing of very heavy character. We believe that this is the only known case of abrasion at such a depth, and it is important to those who study the geography of the seas, inasmuch as the chafe of the cable indicates the existence of a powerful ocean current at a depth of 3,000 feet along the Spanish or Portuguese coast."

MARCH 1871

"Attempts to establish a ready communication between the beleaguered inhabitants of Paris and their relatives and friends beyond the German lines have given rise to contrivances that may introduce a new era in both aeronautics and photography. Among them may be mentioned the ingenious device by which two whole pages of *The Times* of London have been transmitted from London to Paris. The pages were photographed with great care on pieces of almost transparent paper an inch and a half in length by an inch in width. On these tiny scraps there could be seen by the naked eye only two legible words, *The Times*, and six narrow brown bands representing the six columns of printed matter forming a page of the newspaper. Under the microscope, however, the brown bands became legible, and every line of the newspaper was found to have been distinctly copied with the greatest clearness. The success of this experiment gives rise to the hope that the new art of compressing printed matter into a small compass will not stop here."

MARCH 1873

"Mr. Willoughby Smith has been making a series of electrical experiments with selenium. Sticks of selenium were connected with platinum wire and hermetically sealed in glass tubes. The electrical resistance of some of the sticks was very great and the resistance of other sticks was much less. He was at a loss to account for this lack of constancy until after various trials he found that it was due to the action of light. When the sticks of selenium were shut up in a box so as to exclude light, the electrical resistance was highest and remained constant, but when the cover was with-drawn and light was allowed to fall on the sticks, the electrical resistance diminished according to the intensity of the light. These very singular observations may lead to new and useful discoveries."

OCTOBER 1873

"The good news comes to us from Dundee, Scotland, of the safe arrival there of all the remaining survivors of the Hall arctic expedition, consisting of Captain Sidney O. Buddington and 12 others. The incidents and results of this latest polar exploration can be briefly summed up as follows: On the 29th of June, 1871, the steamer *Polaris*, Captain Charles F. Hall, sailed from New York on a voyage of arctic exploration. In August, 1871, she reached latitude 82°16', the highest point ever attained by any vessel. Soon after the ship went into winter quarters Captain Hall was taken ill, and he died on November 8, 1871. Captain Buddington became master. In August, 1872, finding further progress northward impossible, Captain Buddington determined to return home, and the ship started for the south. She was now caught in the ice, and Captain Buddington caused a portion of the provisions and a part of the ship's company to be landed on the ice. On the night of October 15, the *Polaris* broke away from her icy moorings, leaving the hapless party of 19 persons on the ice. On the 30th of April, 1873, after 6 months of dreary drifting, they were rescued and were safely landed at St. John's, New-foundland."

NOVEMBER 1873

"The yearly rings, shown when the trunk of a tree is transversely divided, by which, as is well known, the age of the tree can be determined, do not diminish in relative thickness by a constant law. M. Charles Gros seeks a cause for the irregularity, and he has arrived at the conclusion that the data, mean and extreme, of meteorological phenomena, when known and tabulated, might be compared year by year with the annual ligneous layers formed during such periods in many different varieties of trees. From the comparison it is not impossible that some interesting ideas relating to the laws of development of trees might be obtained. Moreover, once these laws were established, the trees in turn might become precious collections of meteorological evidence for places and times where observations cannot be made."

JANUARY 1874

"Comet IV of 1874, discovered on the 23rd of August by M. Prosper Henry at the Paris Observatory, presents some remarkable peculiarities. Its rapid changes of form, sudden elongation of tail and its brilliancy, which became so great as to render it visible to the naked eye for some time prior to its passage to its perihelion, are considered to be phenomena that may throw light on our hitherto indefinite knowledge regarding the constitution of comets. On the day of its discovery the body appeared in the telescope as a spherical nebulous mass, strongly condensed at the center and exhibiting no traces of a tail. There was little change until August 26, when a tail began to appear. On September 2 the tail had grown to two degrees and continued elongating. The nucleus remained nearly constant in size, although its brilliancy augmented until, on September 10, it became comparable to a star of the fourth magnitude. The spectrum of the comet is found to be composed of three brilliant and very distinct bands. The first is in the yellow portion, the second in the green and the third in the blue.

There was no trace of a continuous spectrum in the intervals between these lines."

FEBRUARY 1874

"Studying the solar spectrum around the beginning of this century, the elder Herschel passed a sensitive thermometer through the successive colors and observed that the greatest heating effect was not coincident with the brightest illumination but at a distance beyond the red, where no rays were visible. The inference he drew from these observations was that the heating rays were distinct from the luminous rays. By the use of photographic papers it was subsequently ascertained that the chemical action of the sun ray appeared to be greater toward the violet end of the spectrum. Hence arose the belief, which the scientific world has generally entertained of late, that the solar radiation was triple in constitution, being separated into a visible spectrum, a heat spectrum and a chemical spectrum. Dr. Draper, in a memoir, points out Herschel's error and also proves conclusively that every part of the spectrum can produce chemical changes. As a consequence the supposed triple constitution of the sun ray must be dropped. There is in the sun ray neither light nor heat nor chemical power as such but simply vibrations, which, when stopped, may manifest themselves in one or another of these phenomena. 'The evolution of heat, the sensation of light, the production of chemical changes are merely effects, manifestations of the motions imparted to ponderable atoms.'"

MAY 1874

"Experiments performed by Professor Arthur W. Wright of Yale College will probably set at rest the moot question as to the nature of the zodiacal light. This faint nebulous radiance is seen in certain seasons, especially in the Tropics, in the west after twilight has ended or in the east before it has begun. Professor Wright, employing a new apparatus of his own design, has determined that the light is polarized in a plane passing through the sun and thus has proved it is reflected and derived originally from the sun. Confirming this view, it was found that the spectrum of the light is the same as that of the sun, except in intensity. Professor Wright concludes that the light is reflected from matter in the solid state, from innumerable small bodies revolving around the sun in orbits, of which more lie in the neighborhood of the ecliptic than near any other plane passing through the sun."

JUNE 1874

"It appears from the soundings made by the *Challenger* expedition that the Gulf Stream, or Florida Current, is a limited river of super-heated water, of which the breadth is about 60 miles near Sandy Hook, N.J., whereas near Halifax, Nova Scotia, it has separated into divergent streams forming a sort of delta. Its depth is nowhere more than 100 fathoms; at less than double that depth we come into what is clearly polar water."

JULY 1874

"Some months ago Darwin wrote to his disciple Fritz Müller, now in Brazil, directing his attention to the habits of the leaf-cutting ants, these ants do not feed on the leaves they gather in such vast quantities but on the fungus which grows on the leaves in their underground chambers. On examining the stomachs of these ants Mr. Müller found no trace of vegetable tissue that might have been derived from the leaves, but only a colorless substance showing under the microscope some minute globules, 'probably the spores of the fungus.'"

AUGUST, 1874

"The natural history of the diamond is one of the puzzles of geology, and until recently the place of its origin was as great a mystery as the manner of its formation. Happily the first part of the problem has been solved: the diamond has been tracked home. Diamonds are found under two very dissimilar conditions: first, as pebbles associated with pebbles of quartz, agate, zoolite and other minerals, and second in circumscribed pits or shafts filled with a chalky or clayey earth, more or less hardened. The shafts are surrounded by a rim of rock: inside the rim, or 'reef,' as the miners call it, the diamonds are found at home and untraveled, whereas outside they occur only in layers of gravel or other products of running water. That the gems within the shaft have rested undisturbed since their formation is attested by the nature of their matrix, which at one South African shaft has been mined to a depth of 200 feet without any apparent decrease in the richness of the yield."

SEPTEMBER 1874

"The French National Assembly has offered a prize equivalent to $60,000 to the inventor of a method for the destruction of *Phylloxera*, a genus of plant lice that attacks grape vines. The *Phylloxera* attack on the vines of France began to attract serious attention soon after the close of our civil war, the roots of the plants affected becoming swollen and bloated and finally wasting away. Professor Planchon, in 1868, recognized the injury as caused by the punctures made by a minute insect. The disease has continued to spread, especially in France, and has now also appeared in Portugal, Austria, Germany and even in England."

DECEMBER 1874

"The new and celebrated painting of the 'Roll Call' is now nightly exhibited in London to large audiences, by means of the oxyhydrogen light, or lime light, and all the colors of the picture are brought out with marvelous brilliancy, in fact with the same perfection as by daylight. The idea of illuminating art galleries in the evening by the lime light is an excellent one. The yellow color of the ordinary gas flame has the effect of revealing only a portion of the colors of the paintings. The reds and yellows are seen well enough, but the blues and greens, and their various tints, are sadly distorted, and the artistic effect lost. The use of the lime light or the electric light obviates such difficulties."

OCTOBER 1875

"One of the most disastrous storms that has ever visited our coast recently swept over a portion of Texas and south-western Louisiana, destroying hundreds of lives and an immense amount of property. Little has been positively determined regarding the cause of these cyclones. From observation it appears they may originate wherever a lower stratum of warm, moist air is rapidly elevated above the sea level. In this moist air an immense mechanical power is stored up, and when condensation caused by its elevation occurs, its moist vapor turns into rain, hail or snow, and an influx of air from all sides rushes in to fill the partial vacuum thus formed. It has been proved that this influx toward a central region is immediately followed by the formation of a whirl, the subsequent development of which is due to further supplies of moist air."

FEBRUARY 1876

"It is a well-known fact that many of the artesian wells sunk for oil, in Pennsylvania and other parts of the country, failed to produce oil but emit great quantities of gas. Some measurements and analyses lately made by Mr. O. Wuth, the well-known chemist of Pittsburgh, go to show that in the gas that is thus passing uselessly into the atmosphere we have a vast store of fuel, and that in allowing it to run to waste we have for years presented the most striking instance of extravagance in the world. The well he measured is thirty-five miles from Pittsburgh. He found the composition of the gas to be C_4H_6, or 80 per cent of carbon and 20 per cent of hydrogen, with but little oxide of carbon and carbonic acid. The well has a tube of five-eighths of an inch diameter, and the pressure of gas is no less than 200 pounds to the square inch. On these data Wuth estimated that the well yielded 1,000,000 cubic feet of gas, weighing 58 tons, *per hour*."

MAY 1876

"According to Professor A. Guyot of Princeton, N.J., the islands of the Pacific are of two kinds, called the lower and the higher. The lower rise but 7, 10 and rarely as high as 100 feet above the level of the sea, while the higher islands reach an elevation of 10,000, 12,000 and even 15,000 feet. There is no transition between them. The most remarkable are the lower islands. Their appearance is very peculiar. The eye is arrested by a white beach; then comes a line of verdure, due to tropical trees; then a lagoon of quiet water of a whitish or a yellowish color, then another line of verdure, and finally, beyond all, the dark blue waves of the ocean. The lagoon inside is but a few fathoms deep, but on the outside of the island the water is 15,000 feet deep. Here then we evidently have a tower-like structure reaching up from the bottom of the sea, and having a depression in its summit."

SEPTEMBER 1876

"Professor Graham Bell has succeeded in transmitting the tones of the human voice by telegraph. The apparatus used by Professor Bell is thus described: Two single-pole electro-magnets, each having a resistance of 10 ohms, were arranged in a circuit with a battery of five carbon elements, the total resistance being about 25 ohms. A drum made of goldbeater's skin, about 2 inches in diameter, was placed in front of each electro-magnet, and a circular piece of clock spring was glued to the middle of the membrane of each drum head. One of these telephones was placed in the experimental room and the other in the basement of an adjoining house. Several familiar questions were, it is said, understood after a few repetitions. The vowel sounds alone are faithfully reproduced; diphthongal sounds and rotund vowels are readily distinguished, but consonants are generally unrecognizable. Now and then, however, a sentence comes out with almost startling distinctness, the consonants as well as the vowels being clearly audible."

DECEMBER 1876

"Although failure has been the fate of every attempt thus far made to navigate the air by mechanical devices, the problem has by no means been given up as hopeless. Certain experiments, made at the expense of the Aeronautical Society in England, to determine the exact lifting pressure of air currents directed against a plane inclined at different angles, have obtained results that are especially promising. The plane used was a steel plate a foot square, and the substitute for the resistance occasioned by the passage of a body at high speed through the air was the blast of a powerful fan blower. The pressure on the plate was 3 lb., indicating a wind velocity of about 25 miles per hour. Inclined at an angle of 15 degrees, the plate felt a lifting pressure amounting to $^1/_2$ lb. The chief thing that remains to be done for the successful solution of the problem of flight is therefore to drive a sufficiently broad-bottomed car at a speed of, say, 40 to 60 miles an hour by means of an apparatus acting on the air. At this velocity the resistance of the air would support the car, at the cost of a relatively small part of the driving force."

JANUARY 1877

"A dispatch from M. Tellier to the French Academy of Sciences announces the arrival of the refrigerator ship *Frigorific* at Pernambuco, Brazil, after a voyage of 70 days from France. The meat transported and kept cool by cold air generated by the Tellier ice machine (which works on the principle of evaporating methylated spirit) was perfectly preserved. It is proposed to load the vessel with meat at La Plata, Argentina, and if the return voyage is successfully accomplished, to establish the regular exportation of South American beef to French markets."

APRIL 1877

"The Central Pacific Railroad Company has lately arranged to have 40,000 trees of the species *Eucalyptus globulus* set out along the 500 miles of the right-of-way of the company. This is only the first installment, as it will require about 800,000 of the trees for the 500 miles of valley where they are to be cultivated. The immediate object of the plan is to increase the humidity of the region and lessen the liability to drought. It is an established fact that the destruction of our forest trees over large tracts of the country is having a direct effect on the climate, and we are glad to know that this company is replacing, at least in part, the forests that have been destroyed."

SEPTEMBER 1877

"Professor Asaph Hall of the Naval Observatory recently announced the interesting discovery of two satellites attendant on the planet Mars. At about 11 o'clock on the night of August 16 Professor Hall, with the aid of the great 26-inch refractor telescope, noticed a very small star following Mars by a few seconds. Three hours later he looked again and to his surprise found that the distance between the planet and the star had not increased, although the former was moving at the rate of 15 seconds per hour. On the morning of August 17 another satellite appeared, and its identity was quickly recognized. The distance of the first satellite from the planet is between 15,000 and 16,000 miles, which is less than that of any other known satellite from its primary. It was exceedingly small, having a diameter of not more than 100 miles. The second satellite is believed to be still closer to the planet and to have a period of less than eight hours. The discovery is a triumph both for Professor Hall and for Mr. Alvan Clark, the maker of the great telescope."

NOVEMBER 1878

"Wilson and Savage, two American missionaries, seem to have been the first, in 1846 and 1847, to bring us information about the gorilla. The skulls and skins sent by Savage have been followed up by complete skeletons and preparations sent by other travelers, and naturalists have been able thus to study the appearance and structure of the most formidable of the man-apes, which is credibly stated to inhabit central Africa from Sierra Leone in the north to Loango in the south. Living in the dense forests of that region and avoiding the presence of man, gorillas are seldom to be met with. We have no authentic portrait of a live adult gorilla. In comparing the skeleton of the adult gorilla with that of man we find that the cranium of the gorilla is very small in proportion to that of man. The contents of the smallest skull of man is given at 62 cubic inches: that of an adult gorilla is given at 34 cubic inches."

DECEMBER 1878

"The discovery of the new metal gallium differs in its history from that of any other element. All other elements have been accidentally discovered, and it has been only after their discovery that we became acquainted with their properties. The 'blende' found at Pierrefitte, in which gallium was discovered by Boisbaudran, was examined by him with a view to finding a new element the existence of which he was firmly convinced. The existence of an element corresponding in its properties with gallium had been predicted in 1871 by Mendelejeff and Newlands. It had been found that the elements form a series of numbers, the amounts and successions of which are governed by certain laws. At the same time it was observed that the properties of the elements showed peculiar relations with the arrangement of these numbers. The series formed by the numbers show irregularities, as if certain numbers were missing. The position to be filled by one of the missing elements indicated the properties of its occupant. These properties had been to the smallest particular described by Mendelejeff. He ascertained the specific gravity of the metal yet to be found as 5.9. In operating on small quantities Boisbaudran found at first 4.7. In 1876, when larger quantities of the metal were obtainable, he found 5,955, exactly as Mendelejeff had predicted."

JANUARY 1879

"Mr. Edison's application for a patent on an electric lamp is before the Commissioner and is taking its regular course. According to the rules of the Patent Office, nothing concerning it can be divulged. It is understood, however, that it is progressing favorably. Mr. Edison has already received seven patents bearing on the electric light and has filed three caveats. Five more similar applications are now under way. He has had a man in the Astor Library searching the French and English patent records and scientific journals, from the earliest dates down to the past fortnight, and says nothing like his arrangement has been revealed. Mr. Edison is making elaborate preparations to introduce an experiment with the electric light. He purposes to commence at Menlo Park with 2,000 lights, using telegraph poles with 15 lights on each arm. This experiment, including the cost of the buildings, engine, generating machine and everything, is estimated at from $100,000 to $125,000."

JANUARY 1880

"The transmission of power by electricity both for short and long distances is not only practicable but also economical; and the sanitary and other advantages of drawing power from a distance, for small manufacturing and for operating domestic machinery, are so enormous that the new system is sure to work great changes in all branches of industrial affairs. It is no stretch of the imagination to say that our children, if not ourselves, will see the small steam engine everywhere displaced by the electric motor, which will convert into motive power the subtle energy conveyed by wires from central sources of energy—huge furnaces constructed on the most approved scientific principles, out-of-the-way waterfalls, tidal currents, even the sun himself. And doubtless this cleanly and trusty servant will serve humanity in ways we are not able to dream of now, and at a cost that will be, by comparison with the present cost of light and heat and working energy, almost nominal."

FEBRUARY 1880

"The vast cavities in the sun we call sunspots are not solid things, and they are not properly to be compared to masses of slag or scoria swimming on a molten surface. They are, rather, rents in that bright cloud surface of the sun we call the photosphere, and through which we look down to lower regions. Their shape may be very rudely likened to a funnel with sides at first slowly sloping (the penumbra) and then suddenly going down into the central darkness (the umbra). This central darkness has itself gradations of shade, and cloud forms may be seen there obscurely glowing with a reddish tinge far down into its depths, but we never see to any solid bottom. We are able now to explain in part that mysterious feature in the sun's rotation, for if the sun be not a solid or a liquid but a mass of glowing vapor, it is evidently possible that one part of it may turn faster than another. Why it so turns no one knows, but the fact that it does is now seen to bear the strongest testimony to the probably gaseous form of the sun throughout its mass."

MAY 1880

"Salicylic acid, the manufacture of which on a large scale has only been rendered possible since 1874 by the patented method of Professor Hermann Kolbe, is the most important antiseptic, antizymotic and antipyretic ever discovered. It is a white, dry, crystalline powder, devoid of smell or taste, and has no detrimental effect whatever on the animal system. Experiments made by the most eminent physicians and surgeons have been so successful that both the acid and its sodium salt are universally recognized as valuable acquisitions in medicine. The scientific papers in all countries have been full of the most favorable reports concerning the immense success obtained in using salicylic acid in treating all cases of gout, neuralgia, acute rheumatism, fevers of all kinds, typhus, cancer, diseases of the throat and stomach, diabetes and other intestinal complaints."

JUNE 1880

"Professor N. A. E. Nordenskjöld's recent grand achievement of navigating the Arctic seas north of Asia, and passing eastward through the Bering Strait down into the Pacific Ocean, has been crowned with a triumphal welcome at his arrival home in Stockholm. The *Vega*, a small steamer built at Bremen, of 300 tons' register, with an engine of 60 horse power, has circumnavigated the entire joint continent of Europe and Asia, having left Gothenburg on July 4, 1878, and arrived at Stockholm April 24, 1880. The voyage consisted of two stages, the first being the passage from the port of departure to that part of the Siberian coast in the neighborhood of the mouth of the river Lena. There, on September 28, the *Vega* became fixed in the ice at a short distance from the mainland, and further progress of the expedition was checked during the winter months. With the return of the brief Arctic summer in July of last year the vessel was released, and Nordenskjöld, boldly pushing his way eastward through unknown waters, succeeded with comparative facility in skirting the coast and, rounding the northern capes of Kamtschatka, was enabled to direct his course southward into the North Pacific Ocean, arriving on September 2, 1879, at Yokohama in Japan."

NOVEMBER 1880

"Some three years ago an intelligent mineralogist discovered specimens of pitchblende on the waste dumps of Denver, Colo., and recognizing the value of the mineral, gathered a quantity and sent it to Swansea, where it brought five shillings a pound, equal to $25 a ton. Pitchblende, or uraninite, is an oxide of uranium, obtained in Saxony and Bohemia, and used in fine glass making. Glass colored with uranium has the peculiar property of showing green when looked at, although perfectly and purely yellow when looked through."

JANUARY 1881

"The island of Madagascar is inhabited only by animals the pedigree of which can be traced back to the Eocene epoch. Madagascar must have been separated from the African continent at least since that time, this being the reason its fauna has developed in a manner quite different from that of other parts of the world."

AUGUST 1881

"Some exceedingly interesting experiments lately took place at the Naval Torpedo Station with the Weeks rocket torpedo. This torpedo is a most peculiar structure. It consists of a float made of tin and sheet iron, braced internally with wood. It has two rudders similar to the tails of a sky rocket. The float portion is some 11 feet long, with the rudders being of the same length. In the forward part or head is placed some 50 pounds of dynamite, and this, coming in contact with an object, explodes by concussion. The entire structure is propelled by a rocket some six inches in diameter, three and a half feet in length and 100 pounds in weight. It moves on the surface of the water and has attained the wonderful speed of about 150 feet per second, which is kept up for some 600 yards. Captain Thomas O. Selfridge, the commandant of the Torpedo Station, witnessed a recent trial. A whizzing noise was heard and the torpedo went on its way. The velocity was something frightful, as may be judged when it is stated that the torpedo passed along and over (for it jumped occasionally) a distance of water not less than 1,375 feet in about *nine seconds*. It was impossible to time it precisely, for the smoke behind was very dense. Captain Selfridge said that the trial was a success."

AUGUST 1882

"A correspondent who has lately gone over the territory devastated by the great fire in the forests of Michigan last fall says his observations are conclusive that phenomena aside from the ordinary conditions of combustion were developed. In the first place the fire created at least two centers that had the essential features of storms, especially the spiral winds. The evidences are confirmatory of the belief that this storm center, after it became fully developed, consisted of a heated body of air or gas in a state of combustion, constantly fed by the smoke and vapor driven to the center by the whirling winds and the gases generated in the combustion of the pines and other resinous woods. This body of air or burning gas by its heat acquired an ascensive force, but by the rapid forward motion of the fire it was sucked forward, actually preceding the fire proper. It is evident that this body was of intense heat, possibly as great as 400° Fahr., at which point oxygen and carbon unite. That such a body of luminous vapor existed, detached from the fire, is asserted by many who saw it from a distance."

NOVEMBER 1882

"The Municipal Council of Paris having voted a grant of 1,000 francs to the Academy of Meteorological Ascension for the purpose of making experiments in aerial photography, a balloon ascent was lately made by members of the academy. They carried with them an apparatus for taking instantaneous photographs. This had six lenses pointing in different directions in order to embrace the whole of the horizon. The balloon rose 200 meters. A telephone was afterward fitted up in the car to enable the occupants to communicate with their friends below."

AUGUST 1883

"The long voyage made in the interest of science a few years since by the *Challenger*, a ship of the British service, awakened a widespread desire upon the part of intelligent people in all portions of the world to learn something further concerning that wide and mysterious domain, the bottom of the sea. A report made by the U.S. steamship *G. S. Blake*, upon her return in February from a somewhat similar mission, has renewed this feeling. The *Blake* is a floating laboratory. She is schooner-rigged and would pass, were it not for certain accessories, for a roomy private yacht. These additions are mainly a heavy boom rigged forward and pivoting upon the foremast, and a high framework over the port bow. The former is used in handling the trawl and the latter is the complex and delicate reel and its belongings by which the miles of wire are paid out or wound in during soundings. Below decks a large proportion of the space is given over to the draughting room, laboratory and storage rooms. Every minute particle brought up during a haul is preserved and labeled. To the ordinary investigator much of this painstaking labor seems wasted, but a little reflection reveals the great utility of all the care. The execution of the work embraced observations of depths, serial water temperatures and densities, and of currents when possible, together with the collection of specimens of the bottom soil and deposit, and of surface, bottom and intermedial water specimens."

APRIL 1884

"Recent research in cerebral physiology has been directed toward the subject of the localization of sensory areas on the cortex of the brain, and it has been productive of many very interesting discoveries. From an analysis of 32 recorded cases with autopsies Dr. M. Allen Starr reaches the following conclusion regarding the pathology of hemianopsis (blindness in half of the visual field): that the visual area lies in the occipital lobe of the cortex, that the symptoms other than the visual cannot be referred to any lesion other than one of the occipital lobe and that the right occipital lobe receives impressions from the right half of both eyes and the left occipital lobe from the left half of both eyes."

OCTOBER 1884

"M. Hervé Mangon has communicated to the French Academy of Sciences a report in which he states that a navigable balloon has at length been perfected by a captain of engineers named Charles Renard. The difficulty was to obtain a motive force in the car of the balloon, the apparatus of which should not be too ponderous for the sustaining power of the balloon itself. Captain Renard discarded the idea of a steam engine, and found the dynamic agent which he sought in electricity, with an apparatus of accumulators, by the force stored in which an engine of ten-horse power could be propelled during several hours. Under these conditions an ascent was made at a recent date. The balloon rose from Meudon and proceeded to Villebon, when, to the astonishment of those watching its progress, it described a semicircle and returned, notwithstanding the apparent opposition of a slight breeze, to the place whence it came. The trial was repeated, with similar results, the aeronaut subsequently declaring that the points where the balloon should halt, and return to its place of departure, had been fixed upon with precision beforehand."

OCTOBER 1885

"At the recent meeting of the American Association, Prof. Langley demonstrated that in addition to the wave lengths in and beyond the visible spectrum, he has detected, by means of the bolometer, vibrations of much greater wave length than have heretofore been known. They are several octaves below the red end of the spectrum. They thus extend the range of recognized vibrations to between six and a half and seven octaves."

OCTOBER 1885

In addition to the value of photography as a means of recreation, it is also found to be an invaluable aid to professional men. Engineers, architects, and draughts-men use it for recording the progress of their work, making pictures of machinery, buildings, copying, drawings, and an infinite variety of work, which saves a vast amount of hand labor. Physicians find it useful in making memoranda of surgical operations. Insurance men use it in inspecting risks and adjusting losses by fire. Artists find it indispensable as an aid to sketching. Correspondents for illustrated papers and magazines now carry cameras as a part of their outfits, and even traveling sign painters photograph their work and send in the picture as a voucher on which to draw their pay.

All this has been accomplished by the introduction of the gelatine dry plate; but there are many who are still prevented from practicing the art by the weight of the apparatus and material, which has to be carried about to make even a few pictures. The weight of the glass is such a serious burden, especially in the larger sizes, that it discourages even the most enthusiastic after a few trials, and many cameras have been laid aside for this reason when they would otherwise be a source of unending satisfaction to their owners.

By reason of several recent improvements it has been found possible to prepare paper of fine and close texture upon a large scale, with an even coating of an extremely sensitive bromide of silver gelatine emulsion, so perfectly that positive silver prints made from the paper negatives will show no grain in the half tones, and be equally as clear and perfect as if made from glass.

The sensitive paper is prepared in sheets for use in ordinary plate holders, and in spools. The film carrier is a flat board made of several strips of narrow wood glued together edgewise to prevent warping, between which and the spring metal frame lying flat, the sensitive sheet is clamped, by laying the board over the back of the paper. The whole is then slid into an ordinary plate holder like a plate.

When used in long strips, the paper is wound upon a wood spool, arranged for use in an instrument termed a roll holder, the principle of which is to draw the sensitive paper from the supply spool at one end over an exposing platform, occupying the same place as the focusing ground glass, to a winding-up reel at the other end.

These parts are enclosed in a highly finished mahogany case, with the vulcanite slide partially withdrawn, exposing to view the sensitive paper lying smooth and flat upon the exposing platform. A removable back supporting the working parts is attached to the case by four flat spring catches.

In taking the holder apart in the dark room to either insert a new spool or remove a reel of exposures, these catches are thrown out, the key socket and indicator knob on the side pulled out, and the case lifted off.

The light, blackened frame of brass holding the various parts is pivoted to the back by two pairs of sliding spring bolts, we draw inward with the right hand the sliding bolts, and with the left raise the whole, this affords access to the spool mechanism. The supply spool is next inserted, by raising the pressure spring brake, and pushing the end of the spool upon a projecting plug, the opposite end being fastened by a thumb-screw. A suitable spring friction mechanism is provided for giving tension to the paper when unwound from spools. After insertion the frame is lowered, locked to the back with the spring bolts, and the opposite end raised. The free end of the paper is then drawn up over an exposing platform and down to the reel under a flat brass pivoted clamp. When the paper is properly centered, the clamp is pressed down, holding and drawing the paper as soon as the reel is rotated. The reel is inserted and removed in a manner similar to the supply spool, but is constructed so that it will be impossible to put it in the same plane as the latter.

A spring pawl bears upon the head of reel and a gravity pawl upon that of the spool; these are thrown off during the process of changing spools.

A guide roll is placed at each end of the platform, the one on the right, being termed a measuring guide roll, in which is a longitudinal slot used as a guide for the point of the knife in cutting off the exposed from the unexposed paper. The roll has a pin at one end for operating a flat spring, making a sound alarm, and in addition on its axis a spur wheel geared with a larger wheel for rotating the indicator. Metal points project slightly above the surface of the reel at each end, which puncture the margin of the paper at each revolution. As the circumference of the roll is one-fourth the length of the picture, four alarms are sounded, and the indicator at the same time makes one revolution, when one exposure has been wound up. A key with a screw thread is inserted in the key valve and rotated like the winding of a clock, revolving the reel, which winds up the exposed sheet, and brings a fresh surface into place ready for the next exposure. When the indicator has made one revolution, and the fourth click been heard, then the operator knows that the change has been accomplished.

After exposure, the paper on the reel may be readily removed, and a new one inserted to take up the balance of the unexposed paper. By counting four dots on the margin from the end after the sheet has been severed, the length of the picture is easily determined, and may be cut off with a pair of scissors.

The exposed sheets, as they are cut off, can be developed several at a time in one tray, with the usual pyro developer; Cooper's developer, being preferred. The developer is sold ready mixed, thereby insuring to the novice success at the outset.

After the negative is fixed and dried, positive silver prints may be made from it in the usual way; but to quicken the process, oiling the paper with castor oil and a hot iron, is recommended, which renders it translucent. Parafine wax may be used in place of oil.

The primary advantage of paper over glass is its extreme lightness. An 8 x 10 apparatus, complete with camera, lens, roll holder for 24 exposures, tripod, and case, weighs 28 pounds less than a glass equipped outfit.

Such a saving makes the taking of large photographs attractive, and enables the amateur to obtain panoramic or other views of inaccessible regions with considerable comfort. The danger of breakage is avoided, thereby making rough transportation of the negatives perfectly safe.

The compact way in which the negatives can be packed should

not be overlooked; they can be kept in books, thereby affording as easy a means of reference as if they were in a photographic album—a point of much value in any large concern. They can be used in photographic ink printing processes without the need of transfer, so common with glass plates. They are splendidly adapted for large work, and, as an instance of their success in this respect, we have but to refer to the very fine exhibition of life-sized direct portraits which was given at the Buffalo Photographers' Convention, in Buffalo, N. Y.

The softness and delicacy of the shadows and the brilliancy of the high lights were specially noticeable.

The retouching of paper negatives is more easily done than on glass, for the back of the negative is worked upon by a pencil; any mistake can be readily erased. With crayon stubs very pretty cloud effects can be worked into the sky of landscape negatives. Perfect freedom from halation is one of the special characteristics of the paper, making it valuable in the photographing of interiors. All portions of the holder are made interchangeable.

The enterprise of the Eastman Company in introducing so noteworthy an invention as their roll holder, and the excellent sensitive paper film used with it, is illustrative of the characteristic push and energy so often displayed by American inventors; we bespeak for their improvement an important future, and consider it an advance in the art of photography which will be welcomed both by the amateur and professional. A silver medal was awarded the company at the London International Inventions Exhibition for the novelty of the invention and the fine workmanship displayed.

NOVEMBER 1885

"We are apt to regard the rain solely as a product of distillation and, as such, very pure. Yet a great number of industrial processes, domestic combustion of coal, natural changes in vegetable and animal matter, terrestrial disturbances such as tornadoes and volcanic eruptions, vital exhalations, etc., are discharged into the atmosphere and, whether by solution or mechanical contact, descend to the surface of the earth in the rain. The acid precipitation around alkali and sulphuric acid works is well known; the acid character of rains collected near and in cities and the remarkable ammoniacal strength of some local rainfalls have been fully discussed."

APRIL 1886

"A new sweetening agent has been produced from coal tar. It is known to chemists as benzoyl sulphuric imide, but it is proposed to name it saccharine. The discoverer is Dr. Fahlberg, and its preparation and properties were recently described by Mr. Ivan Levinstein at a meeting of the Manchester section of the Society of Chemical Industry. Saccharine is about 230 times sweeter than the best cane or beet-root sugar."

JULY 1886

"In these days of remarkable exhibitions of skill in playing baseball by professional exemplars of the game, one cannot look back to the early period without being struck by the great contrast between the work done on the diamond fields at Hoboken, in the 'fifties,' and that which marks the play of the leading professional teams of the present era. The game has been wonderfully improved, and in nothing so much as in the great degree of skill now shown in the pitching department. Modern pitching excels the old method in one special feature, and that is in *the horizontal curve of the ball through the air.*"

AUGUST 1886

The last operation before the Statue of Liberty left France was the assembling of all of the many pieces comprising the shell or statue proper and the final fitting of each piece to each of its surrounding neighbors. Each piece was then marked with a particular number or figure, and every two meeting pieces were designated by the same character marked upon their adjoining edges to serve as a guide when reassembling the statue upon its pedestal at Bedloe's Island in New York. Surrounding each separate piece at a short distance from the edge is a row of small holes; when two pieces are joined together, the holes in one coincide with those in the other, so that the two may be firmly united together by rivets.

The copper of the shell, being only about three thirty-seconds of an inch thick, lacks rigidity, so that it was necessary to increase the stiffness of every piece by means of iron bars secured to the interior surface. These bars are so disposed and united to each other as to form a most intricate network of bracing, covering and strengthening the entire statue.

Lightning has several times struck the ironwork, but, owing to the means that were early taken to lead the current away, not the slightest damage has been done. Up to the present time, no portion of the foundation has settled; and the solid concrete foundation proper which is easily the largest single block of artificial stone in the world, being ninety feet square at the base, sixty-five feet square at the top, and fifty-two feet ten inches in height, with a central well hole ten feet square, is without crack or flaw of any description.

FEBRUARY 1887

"Salt is now the means for clearing away snow in Paris. A regular service for the removal of snow on its first appearance has been organized, as it is important to clear away the snow before it has been compressed into ice by the passage of vehicles, when it is far more difficult to remove. As falls of snow rarely occur at Paris with a temperature much below the freezing point, salt may be sprinkled on the snow, producing a liquid, of which the temperature may descend to 5 deg. Fahrenheit without its freezing. This cold mixture does no harm to paved roads, asphalt and wood pavements. The small cost of the system, and the advantages to traffic, are sufficient reasons for an early and wide extension of this use of salt. In this work the street crews spread salt that is not suitable for ordinary purposes."

JUNE 1887

"Those who have been kept from the pleasures of 'cycle riding by fear of accident on the high wheel will dismiss all their fears if they use the Rudge bicyclette. The wheels are of equal size. The front wheel is the steerer, so that the power required to propel the machine does not affect the steering, as in the ordinary bicycle."

JANUARY 1888

"The mastodon, that great fossil mammal, allied somewhat nearly to the elephant, has become perhaps more familiar to the public than any other of the numerous great creatures that once lived in our extended country. In nearly every state west of New England portions of this creature have been disinterred. The circumstance of several examples having about them evidences of man's work is extremely interesting. We are therefore able to say that man and mastodon are contemporaneous. But the date is obscure. We have not determined what sort of man made the stone arrowheads that struck the life out from the great carcasses and lie among their remains."

NOVEMBER 1888

"If ships could be pulled through the water, instead of being driven by means of the ordinary propellers now in use, then a saving, on the average, of 40 per cent, in engine power would be gained. Both the screw and the paddle push the sustaining water from the ship, and thus augment its resistance, and it has been found by experiment that the thrust of the propeller surpasses the pull of a tow rope, on the average, by 40 percent, to obtain the same speed in the same vessel. The solution is to propel the ship by means of revolving sails acting in the air. The air propeller is, in its outer shape, somewhat similar to the ordinary water screw, with sails or blades made of thin sheet steel."

FEBRUARY 1888

"The distances of the stars are ascertained in the same manner as those of the sun and planets; that is, by parallax. Instead, however, of taking two stations at different parts of the earth's surface and laying down a base line between them, we take the diameter of the earth's orbit, or 183,000,000 miles, as the base, the observations being made at intervals of six months. Even with this immense line, however, the parallax is so small that it can only be detected by the most careful observations and accurate instruments. The star Alpha Centauri is closest to the earth."

JUNE 1889

"At the Eiffel tower, an experiment was performed recently which produced a strong impression on those present. The engineer of the American firm of Otis subjected the Otis lift to a final test before handing it over for public use. The lift was fastened with ordinary ropes, and this done it was detached from the cables of steel wire with which it is worked. What was to be done was to allow the lift to fall, so as to ascertain whether, if the steel cables were to give way, the brakes would work properly and support the lift. Two carpenters armed with great hatchets ascended to the lift; at a given signal, a blow cut the rope and the enormous machine began to fall. Every one was startled, but in its downward course the lift began to move more slowly, it swayed for a moment from left to right, stuck on the brake, and stopped. There was a general cheering. Not a pane of glass in the lift had been broken or cracked, and the car stopped without shock at a height of ten meters above the ground."

JULY 1889

"Professor John Trowbridge calls attention to the importance, from an engineering point of view, of making careful photographs of steel and timber at the point of rupture under a breaking load, suggesting that in this way we may learn something important on the much vexed question of elasticity."

DECEMBER 1889

"Great cats with enormous canine teeth projecting from the upper jaw formerly roamed the earth, but have disappeared, leaving not a single analogous representative. These great-toothed cats have been called saber-toothed tigers (*Smilodon and Machaerodus*). They must have used their great teeth as daggers for ripping and thrusting instead of biting, for it was evidently impossible for them to open their jaws wide enough to free the lower jaw from the tips of the great upper canine teeth."

MARCH 1890

"Aluminum, whether pure or in combination, deserves to rank with the noble metals. Prior to 1887, the entire amount of aluminum manufactured annually was only 10,000 pounds, and it sold that year at $10 a pound. The Hall process, on patents of Charles M. Hall, is now being carried on by the Pittsburgh Reduction Company, who are now selling pure aluminum at a rate cheaper than nickel. Briefly, Hall's process is this: A flux being discovered that, at a moderate temperature, takes the aluminum ore into solution, that is of lighter specific gravity, and that also is unaffected by the passage of an electric current, he fills a series of carbon-lined steel pots with the flux, which is kept molten. Carbon electrodes are plunged into these baths, through which passes the electric current, which acts to send the aluminum to the sides and bottom of each pot."

JULY 1890

"A novel telephone station is being introduced in Connecticut. The instrument cannot be used unless a fee is paid. If five cents is dropped in the slot, it strikes a bell of a high note, once. A quarter strikes a bell of a lower note, once. A half dollar strikes that bell twice, while a silver dollar strikes a very low tone 'cathedral gong.'"

SEPTEMBER 1890

"Up to a recent period heat was always supposed to be a prime factor in the rolling of steel, and that without it no alteration in what may be styled granulation was possible. Now a Chicago paper announces a change in manipulation that completely explodes the old theory. Bars of cold steel are as easily rolled into wire as if the metal were hot, and not only that, but the process nearly doubles the tensile strength.

Manifestly, if wire can be rolled from cold bars with such results, why may not steel plates for ships or other purposes? Yea, why not even railroad bars? The ending of a beginning in what is new now is beyond the ken of the wisest."

OCTOBER, 1890

"Let us see what might be the consequences of a celestial meeting of the earth (traveling 18 miles per second) and a comet that had at least an equal velocity. If the comet had a consistent nucleus, the terrestrial crust would be staved in by the impact, and the torrents of lava that it conceals would produce a terrible commotion in contact with the waters of the ocean. In addition, the axis of the earth would be abruptly displaced. This is the sole plausible hypothesis to explain the inclinations of planets upon their orbit."

CHAPTER THREE

FEBRUARY 1891

"If a rapid vertical fall assumes an exceptional character of magnitude, it will produce physiological disturbances of the same kind as those that a person experiences in rustic swings, toboggan slides, merry-go-rounds, the sight of abysses, etc. Such is the field to be exploited. A tower several hundred meters in height and a closed cage constitute the plant. The passengers enter the cage, which is allowed to drop freely from the top of the tower. At the end of 100 meters fall the velocity acquired is 45 meters per second, and at the end of 300 meters it is 77 meters. In order to render this maneuver practical, it suffices to receive the passengers safe and sound at the end of the trip. This object may be realized by giving the car the form of a shell with a very long tapering point, and by receiving it in a well of water of sufficient depth. The accompanying figures give the general aspect of such a shell."

AUGUST 1891

"The banana belongs to the lily family, and is a developed tropical lily, from which, by ages of cultivation, the seeds have been eliminated and the fruit, for which it was cultivated, greatly expanded. In relation to the bearing qualities of this fruit, Humboldt said that the ground that would grow 90 pounds of potatoes would also grow 33 pounds of wheat, but that the same ground would grow 4,000 pounds of bananas."

OCTOBER 1891

"The widespread sails of a ship, when rendered concave by a gentle breeze, are most excellent conductors of sound. A ship was once sailing along the coast of Brazil, far out of sight of land. Suddenly several of the crew noticed that when passing and repassing a particular spot they always heard with great distinctness the sound of bells chiming sweet music, as though being rung but a short distance away. Several months afterward, upon returning to Brazil, the sailors were informed that at the time when the sounds were heard, the bells in the cathedral of San Salvador, on the coast, had been ringing to celebrate a feast held in honor of one of the saints. Their sound, favored by a gentle, steady breeze, had traveled a distance upward of 100 miles over the smooth water, and had been brought to a focus by the sails."

DECEMBER 1891

"M. de Chauveau had the idea of exporting water from the Dead Sea as an antiseptic for use in hospitals, it being reputed mortal to every kind of animal life, and necessarily, as he supposed, to microbes. But a savant whom he consulted said, 'Take care, there is hardly a fluid in nature in which a virulent microbe of some sort may not find a good soil.' He therefore cultivated various kinds of bacilli in the densest Dead Sea water that had ever been fetched to his laboratory. The diphtheria, measles, scarlatina, small pox, and other fell creatures of the animalcular world were experimented upon. All died but two, with which in forty-eight hours the fluid was alive. The one shaped like the clapper of a bell and the other like a tack nail with a round head were the microbes of tetanus and of gangrene."

JANUARY 1892

"Inhabitants of cities indulge far too freely in meat, often badly cooked and kept too long; the poor and country population do not often get their meat fresh. Professor Verneuil considers something should be done to remedy this state of things. He points out that Reclus, the French geographer, has proved that cancer is most frequent among those branches of the human race where carnivorous habits prevail."

FEBRUARY 1892

"Henry Wilde, Fellow of the Royal Society, has come to the conclusion that the outer shell of the earth and the great mass within rotate somewhat independently of each other. The interior portion, still in a liquid condition, he conceives as continuing to revolve about the axis which our planet had in its infancy. Somehow, in the great cataclysm in which the moon was thrown off from the earth, the crust of our globe was, he thinks, skewed over to one side about twenty-three degrees. Mr. Wilde constructs a machine, consisting of one sphere within another slightly larger one, both converted into magnets by coils of wire encircling them. Upon those portions of the shell which correspond to the oceans he attaches magnetized sheet iron. He makes the inner and outer spheres rotate on axes 23 degrees apart. Finally, for test purposes, he provides for temporarily fixing a magnetic needle at any point on the surface of the globe. With this ingenious apparatus, he declares he can reproduce every known variation of intensity and direction in terrestrial magnetism of which he can find a record; and, what is the convincing feature of his experiment, the real magnetic history of all parts of the world for the last four centuries, so far as he can learn it, is actually repeated in the minutest details when the inner sphere is made to fall behind the outer one, in their revolution, at the rate of 22 minutes of an arc annually!"

JULY 1892

"A pound of coal used to make steam for a fairly efficient refrigerating machine can produce an actual cooling effect equal to that produced by the melting of 16 to 46 pounds of ice. These figures are sufficient to prove the practicability of artificial cooling for office buildings, hospitals, theaters, hotels, and even for the best class of private houses. It is a curious example of the slowness with which people take advantage of modern inventions that thousands of men sit sweltering in hot offices in the midsummer days; business lags and the efficiency of workers is greatly reduced. At the same time not more than three or four blocks away are great provision warehouses where the temperature is kept at freezing the year round. If it pays to keep dead ducks and turkeys cool on Greenwich Street, why would it not pay to keep live business men cool on Broadway?"

SEPTEMBER 1892

"In order to test the power of gripping in the young infant, Dr. Louis Robinson placed his fingers against the palm. The contact at once caused the hand to close apparently by pure reflex action, since it made little difference whether the child was asleep or awake. He then slowly, but with a slight jerking motion, lifted his fingers, and found to his surprise that the child tightened its grasp and allowed him to raise it from the bed. In many cases a newly-born child would hang and support its weight with ease for a minute. Still more surprising was the fact that in most instances it would make no objection to the experiment whatever. Among the newly-born offspring of the human race these faculties were of no use. It seemed, therefore, legitimate to infer that the astonishing prehensile power in the hands of the modern infant was a vestige of the habits which for many epochs saved their arboreal forefathers during their tender youth from destruction."

FEBRUARY 1893

"The enormous strides made by electricity in commerce and industries have been, to a certain extent, paralleled by applications in medicine and surgery. One of the new features of electric medication is the introduction of drugs into the human body through the skin. This is done by placing solutions of any drug upon a sponge, which is made the positive pole and placed against the skin. When the current is turned on, the drug is actually driven through the skin into the tissues. The application is not at all painful. Thus cocaine has been driven in over a painful nerve, and neuralgias have been relieved by it. Many other drugs have been used in this way. This property of electricity is known as cataphoresis. Operations have been performed after anaesthetizing the skin and subjacent tissues cataphoretically."

APRIL 1893

This is pre-eminently the age of athletics. Within the past twenty-five or thirty years a very remarkable revival in athletics has taken place in this country.

The Racquet and Tennis Club is a luxurious home where the members may shut out the busy world, don their flannels, and after an hour or more of such form of active exercise may, if tired and exhausted, enjoy the delightful lassitude of a Turkish bath, or, he may take a plunge in the capacious swimming tank. Then a half hour on a divan with, perhaps, a cooling beverage at his elbow, our refreshed athlete is ready to stand on the scales and find how much his exercise has reduced his weight.

On the first floor are the pool and billiard rooms, the dining room, two reading rooms, and a reception room. In the basement will be found the bowling alleys and admirable shooting galleries. Also the plunge and the Turkish and Roman bath rooms, all fitted up in white marble and tile.

On the second floor is the lounging room, where the members usually sit while waiting for their turn to secure a court. At the left is the card room and at the right the dressing alcoves, and at the extreme end are the shower and needle baths. The visitor will find on the next floor a large and completely appointed gymnasium, sparring and fencing rooms, and the barber shop.

On the top floor will be found perhaps the most interesting feature of the club–the tennis court. Tennis is a comparatively new game in this country and the court is the first and only one ever built in New York.

MAY 1893

"From the experiments recently performed in electrical oscillations, the conclusion that light and electrical oscillations are identical is very strongly substantiated. The principal parts in which they practically agree are the velocity, rectilinear propagation, laws of reflection, interference, refraction, polarization and absorption by material substances. In fact, the sole certain difference appears to be the wave length. In the domain of wireless telegraphy this subject is of prime importance. Although existing methods are far from perfect, we can confidently expect that in the near future we will be able to telegraph on land and sea without wires by means of electrical oscillations of high power and frequency."

JULY 1893

THE WONDERFUL "MERRY-GO-ROUND" designed by engineer George W. G. Ferris, of Pittsburg, Pa., (sic.) is now completed and forms a most remarkable and attractive object. The inauguration of the great Ferris Wheel took place on the 21st of June, and was a very happy affair. A large number of guests were invited, speeches were made by several distinguished persons, and many compliments were showered upon the engineer and projector of the wonderful machine.

This curious piece of mechanism carries thirty-six pendulum cars, each seating forty passengers; thus one revolution of the wheel carries 1,440 people to a height of 250 feet in the air, giving to each passenger a magnificent view and a sensation of elevation akin to that of a balloon ascent. On the day the wheel was first started 5,000 guests were present at the inaugural ceremonies, all of whom were given a ride on the great wheel. The motion of the machinery is said to be imperceptible.

The 36 carriages of the great wheel are hung on its periphery at equal intervals. Each car is 27 feet long, 13 feet wide, and 9 feet high. It has a heavy frame of iron, but is covered externally in wood. It contains forty revolving chairs, made of wire and screwed to the floor. To avoid accidents from panics and to prevent insane people from jumping out, the windows will be covered with an iron grating.

The charge for a ride in the novel machine is 50 cents, for which the passengers enjoy two revolutions, occupying half an hour. If all the seats are full, the company takes in $1,440 an hour. It is truly a wheel of fortune for its owners.

DECEMBER 1893

"What Sir R. Ball has to say concerning the movements of the molecules in a diamond is truly surprising. Every body is composed of extremely, but not infinitely, small molecules. Were the sensibility of our eyes increased so as to make them a few million times more powerful, it would be seen that the diamond atoms are each in a condition of rapid movement of the most complex description. Each molecule would be seen swinging to and fro with the utmost violence among the neighboring molecules and quivering from the shocks it receives from the vehement encounters with other molecules."

JANUARY 1894

"That the continent of Europe is passing through a cold period has been pointed out by M. Flammarion, the French astronomer. During the past six years the mean temperature of Paris has been about two degrees below the normal, and Great Britain, Belgium, Spain, Italy, Austria, and Germany have also been growing cold. The change seems to have been in progress in France for a long time, the growth of the vine having been forced far southward since the thirteenth century; and a similar cooling has been observed as far away as Rio de Janeiro."

JANUARY 1894

The beautiful peristyle which excited the admiration of millions during the summer has passed away in smoke; but, thanks to photography, its image exists and may be projected in a realistic way upon the screen, still an object of wonder and admiration to both those who have seen the original and those who have not.

One of the uses of the optical lantern is to perpetuate beautiful forms and scenes, thus cultivating the taste and increasing the knowledge of multitudes who could never have seen the object themselves. With greater perfection comes greater usefulness, so that improvements in photography, in conjunction with improvements in the optical lantern, yield results which a few years ago were not regarded as among the possibilities. To perform satisfactorily the great variety of demonstrations possible by means of optical projection, the optical system should be accurately constructed, and the various parts of the apparatus should be interchangeable throughout and so easily adjustable that no inconvenience may be experienced in obtaining the desired results.

JUNE 1894

The junction of England with the continent of Europe has already been the subject of numerous projects. Without going back to the project for a subterranean route recommended in 1892 by Mr. Mathieu, engineer of mines, it will suffice to recall the project for a submarine tunnel proposed by Mr. Watkins, and for a gigantic bridge, whose promoters were Messrs. Schneider and Hersent, as well as the modification of the latter proposed by Mr. Bunau-Varilla.

All these projects have been abandoned, as much on account of the objections urged against them as by reason of the incomprehensible hostility that the English have always manifested toward all enterprises of this kind.

But Sir Edward Reed, a member of the English Parliament, former Lord of the Treasury and engineer in chief of the Admiralty, has taken up the question again with a project which has been received with favor by a large number of members of Parliament, and which therefore seems to have serious chances of success, and the more so in that it avoids the difficulties and objections that were urged against its predecessors.

The project consists in simply submerging, between a point of the French coast situated in the vicinity of Cape Gris-Nez and another on the English coast located between Dover and Folkestone, two tubes that would constitute two absolutely separate tunnels, each serving for the passage in one direction of trains drawn by electric locomotives.

The tube would be of steel plate with double walls and the intervening space would be re-enforced by I beams and filled in with concrete. The putting in place would be effected by sections of 300 feet, hermetically sealed at each end and floated to the place where they were to be submerged.

One of the extremities of the section having been fixed upon a sort of caisson that will afterward perform the functions of a pier, the caisson is weighted so as to cause it to sink. The other extremity continues to emerge, and receives the end of the following section, the junction being made by huge hinges. The caisson of this section is sunk, and so on. When all the sections are in place, the formation of the joints is begun.

What we have said about one tube applies also to the other, but, in reality, Sir Edward prefers to sink the sections of the two tubes simultaneously in properly cross-bracing them, in order to form a sort of rigid girder that would present much greater resistance to transverse stresses.

The caissons forming piers are designed to support the tubes at a slight distance from the bottom of the sea. This arrangement possesses the double advantage of doing away with any preliminary dredging, since it will be possible to give the piers the height necessary to avoid the slight changes of level of the bottom and of assuring a free circulation of the marine currents beneath as well as above the tubes. It permits, besides, of so regulating the system that the upward thrust partially balances the weight of the trains in each section. The stresses to which the tube will be submitted by the fact of such passage will be diminished by so much, and, therefore, much better conditions of resistance will be obtained than in an ordinary bridge.

The use of two distinct tubes will prevent all chances of accidents and will have the great advantage of realizing the important problem of the aeration of the tunnel, without any expense and in as satisfactory a manner as possible. In fact, each train will have somewhat the effect of a piston that forces the vitiated air before it and sucks in pure air behind it to take the place of the former.

The total cost of the installation of the tubes is estimated by Mr. Reed at seventy-five million dollars, which is less than half the cost anticipated by Messrs. Schneider and Hersent for the construction of a bridge across the channel.

SEPTEMBER 1894

"At a recent meeting of the Chemical Society, at the Royal Institution, Professor Dewar gave an account of the researches he has lately been carrying out in connection with the behavior of substances exposed to light at a temperature of 180° below zero. The professor experimented with various definite organic compounds. A hydrocarbon like paraffin was feebly phosphorescent at ordinary temperatures, at low ones, it was brilliantly so. Many of the complex compounds known as ketones were exceedingly luminous—acetophenone, for instance."

NOVEMBER 1894

"The whole world owes a deep debt of gratitude to the young French savant, Dr. Roux, for the discovery of an effectual cure for diphtheria. Diphtheria is produced by microbes which plant themselves in the membrane of the throat, and multiply. There, they secrete a poison of extreme violence, called 'toxin,' which quickly penetrates the circulation and infects the whole body. Dr. Roux's 'serum therapy' is produced by first injecting isolated toxin into a horse. The second step is to draw from the animal a judicious quantity of blood. If the blood be allowed to stand for a while, the red corpuscles settle to the bottom, and the operator can draw off the fluid containing the serum, or antitoxin. This, in turn, is injected under the skin of a patient. The distinguished Dr. Marsan says there are toxins and antitoxins for all microbic affections. Serum therapy will eventually discover a remedy for all infectious diseases."

MARCH 1895

"Lord Rayleigh startled the world by announcing the discovery of a new constituent of the atmosphere. The new gas is called 'argon'; and, so far as is at present known, it stands entirely unrelated to any other chemical substance in nature."

JANUARY 1896

"N. A. Langley has succeeded in obtaining helium perfectly free from nitrogen, argon, and hydrogen. This gas, when weighed, proves to be exactly twice as heavy as hydrogen, the usual standard. Guided by purely physical considerations, the experimenter arrived at the conclusion that the molecule of helium contains only one atom. Hence the atomic weight must be taken as 4."

MARCH 1896

"The theory that the two cerebral hemispheres are capable, to some extent, of independent activity, has been evoked to account for those strange cases in which an individual appears to possess two states of consciousness, such cases as afford the basis of fact for Robert Louis Stevenson's weird romance of 'Dr. Jekyll and Mr. Hyde.' Dr. Lewis C. Bruce records a case which is strongly in favor of the double brain theory. An inmate of the Derby Borough Asylum was a Welshman by birth, and a sailor by occupation. His mental characteristics had different stages at different times. In an intermediate stage he was ambidextrous, and spoke a mixture of English and Welsh, understanding both languages; but he was right handed while in the English stage and left handed in the Welsh stage."

APRIL 1896

"The overground power plant at Niagara Falls is already regarded as one of the local attractions of Niagara. But the casual visitor fails to see the best of the work. Out of his sight below the solid floor, and directly beneath the dynamos, a great rectangular pit descends nearly two hundred feet through the solid rock. Near the bottom, the power company has installed great turbine water wheels, from each of which a vertical shaft rises to ground level to directly drive the rotating fields of the 5,000 H.P. alternators. The station now appears as a purveyor of electric energy, while originally it was intended rather to sell hydraulic power."

MAY 1896

"Each year the laws of sea storms are understood more perfectly through the indefatigable efforts of the United States hydrographic office. The landsman hardly appreciates what has been done by the government to protect ships from danger. In order to measure the storms, it was necessary to obtain reliable data from a wide extent of ocean territory. In the absence of telegraph stations, forms for keeping observations were issued to every captain of a vessel touching any

American port, to be filled out and mailed to the headquarters at Washington. In return for this labor every captain received free the Monthly Pilot Chart. From the pile of data received, a map of each storm was constructed, and rules were compiled that are given to mariners when encountering a storm at sea."

SEPTEMBER 1896

"The United States Patent Office is ready to grant patents for medicines, although it is an open question in professional ethics whether a physician should patent a remedy. Synthetic medicines, prepared by chemical processes, often coal tar products, are now invading the field of Nature's simples, and it is possible that there may yet be a number of patentable medical compounds invented, to replace quinine and other vegetable alkaloids and extracts."

FEBRUARY 1897

"In New York a heavy snow storm is the signal for the marshaling of all the forces of the Department of Street Cleaning. For days a solid procession of carts, filled with snow, is seen in progress down the side streets toward the river, where it is dumped. There have been many experiments directed toward the elimination of the bulky material by some less clumsy and expensive method. A naphtha-burning snow melter was recently tested in New York. The flame of the naphtha and air comes into direct contact with the snow, melting it instantly. Fourteen men are necessary to feed the insatiable monster."

OCTOBER 1897

"The latest Arctic adventure of Lieut. R. E. Peary, U.S.N., while devoid of sensational adventures and discoveries, was crowned with success from a scientific point of view. The great meteorite and the collections he gathered are worth all the expense and labor of the voyage. His vessel the Hope came into Sydney, Cape Breton, on September 20, nearly as deep in the water as when she left the port for the North—the great Cape York meteorite, the largest in the world, being in the hold embedded in tons of ballast. The meteorite is estimated to weigh up to 90 tons, and is composed of about 92 per cent iron and 8 per cent nickel."

MARCH 1898

"An advance as important as the introduction of the internal combustion motor has been made by Mr. Rudolph Diesel, of Munich. The experiments which led to the construction of the present successful machine began in 1882. In the ordinary gas or oil engine, the charge within the cylinder is ignited by a jet, hot tube or electric spark. In the Diesel motor the temperature of ignition is secured by the compression of pure air. Air is compressed to a pressure of up to 600 pounds to the square inch and the fuel, kerosene, is injected gradually into the cylinder and is burnt steadily during the stroke of the piston."

MAY 1898

"The War Department has prepared a system for identifying the men in the United States armies who may go into action. They will wear around their necks little tags of aluminum, by which they may be identified if found on the field of battle. In the last war it was often impossible to properly identify the dead soldiers, and thousands were buried in graves marked 'unidentified.'"

JUNE 1898

"Prof. Dewar has recently liquefied hydrogen, which is an unprecedented feat. Fuller accounts of his experiments have now been published. In the apparatus used in these experiments, hydrogen was cooled to -205° Centigrade, and under a pressure of 180 atmospheres, escaped continuously from the nozzle of a coil of pipe at the rate of about 10 or 15 cubic feet per minute, in a vacuum vessel. Liquid hydrogen began to drop from this vacuum vessel into another, and in about five minutes 20 cubic centimeters were collected."

AUGUST 1898

"The new Zeiss binocular field glasses, now being manufactured in this country by the Bausch & Lomb Optical Company, Rochester, N.Y., are the invention of Prof. Ernst Abbe, of Jena, to whom optical science owes so many recent improvements. The three principal defects of the ordinary field glass are overcome by the use of two pairs of prisms, which erect the inverted image formed by the object glass, shorten the telescope by two-thirds and place the object glasses farther apart than the eyepieces are, thus increasing the stereoscopic effect."

SEPTEMBER 1898

"At the Cambridge Congress of Zoology Prof. Ernst Haeckel read a fascinating paper on the descent of man. He does not hesitate to say that science has now definitely established the certainty that man has descended through various stages of evolution from the lowest form of animal life, during a period of a thousand million years. 'The most important fact is that man is a primate, and that all primates—lemurs, monkeys, anthropoid apes, and man—descended from one common stem. Looking forward to the twentieth century, I am convinced it will universally accept our theory of descent.'"

MARCH 1899

"The great observatories of the world are near large cities or universities—places selected from local or political motives—where atmospheric conditions make them unfit for the most delicate astronomical research. It was a bold step to deviate from this precedent, but this step was taken, and taken by a woman, Miss Catherine Bruce, of New York, who gave $50,000 to the Harvard College Observatory. The Bruce photographic telescope, of 24 inches aperture, is mounted in Arequipa, Peru, in a climate unsurpassed for astronomical work. By its aid, new stars have been found in the Large Magellanic Cloud, showing an additional connection of this object with the Milky Way."

MAY 1899

"At some period in the future a successful substitute for coal may be discovered, but we must bear in mind the extreme cheapness of coal and the possibility of further economizing its consumption. If during the next half century the nation [Britain] is spared international difficulties, such as a great war, we may expect to enjoy a most prosperous period in our manufacturing industries. Eventually, as coal becomes dearer in this country, manufacturing operations that supply the world will gradually be transferred to countries where the cheapest coal is produced."

JUNE 1900

"An ocean depth of 5,260 fathoms, or 31,560 feet, has been found by the United States steamer 'Nero,' which has lately been engaged in making soundings between Guam and Manila. In November, 1899, the 'Nero' reported a sounding of 4,900 fathoms about 500 miles east of Guam. The deepest ocean sounding heretofore reported was 30,930 feet, northeast of New Zealand in the South Pacific."

MAY 1901

"If Prof. J. J. Thomson's corpuscular hypothesis be absolutely demonstrated, our ideas in regard to chemistry will be revolutionized. Prof. Thomson concluded that the small particles carrying the charges of electricity were only one-thousandth of the size of an atom. These experiments were all made with discharges of negative electricity. It was also found that these small particles negatively charged were given off from incandescent matter and from radium. When he first enumerated his theory to the scientific world three or four years ago, it was received with considerable incredulity, but has now been adopted by many scientists. He regards the chemical atom as made up of a large number of similar bodies which he calls 'corpuscles.' A normal atom forms a system which is electrically neutral. The electrification of a gas consists in the breaking off from the atoms of a few corpuscles."

JULY 1901

"One hundred years ago it was generally believed to be impossible for two substances of entirely different properties to have the same composition. This phenomenon of isomerism, so rare at one time, is now very common. We have, for example, 55 substances having the formula $C_9H_{10}O_3$, all having the same elements in the same proportions, or the same kind of atoms and the same number of atoms of each kind. To explain isomerism it was necessary to assume that in these different bodies the atoms are differently arranged or grouped. Since 1888 a great deal of work has been done in the development of the theories of space chemistry or stereochemistry. We are in a position now not only to determine how the atoms are linked to one another, but also how they are actually grouped in space. Stereochemistry is the most attractive field of research in organic chemistry today. Prominent among the men who have contributed to this department of chemistry are van't Hoff, Wislicenus, Baeyer, and Emil Fischer."

SEPTEMBER 1901

"H. Becquerel has confirmed, by an unpleasant experience, the fact first noted by Walkoff and Giesel, that the rays of radium have an energetic and peculiar action on the skin. Having carried in his waistcoat pocket for several periods, equal in all to about six hours, a cardboard box enclosing a small sealed tube containing a few decigrammes of intensely active radiferous barium chloride, in ten days' time a red mark corresponding to this tube was apparent on the skin; inflammation followed, the skin peeled off and left a suppurating sore which did not heal for a month. A second burn subsequently appeared in a place corresponding to the opposite corner of the pocket where the tube had been carried on another occasion. P. Curie has had the same experience after exposing his arm for a longer period to a less active specimen. The reddening of the skin at first apparent gradually assumed the character of a burn; after desquamation a persistent suppurating sore was left which was not healed fifty-six days after the exposure. In addition to these severe 'burns' the experimenters find that their hands, exposed to the rays in the course of their investigations, have a tendency to desquamate, the tips of the fingers which have held tubes or capsules containing very active radiferous material often become hard and painful; in one case the inflammation lasted for fifteen days and ended by the loss of the skin; and the painful sensation has not yet disappeared after the lapse of two months."

NOVEMBER 1901

"An Italian engineer. M. Triulzi, has devised a special instrument, the cleptoscope, whereby it is possible for the crew of a submarine boat to ascertain what is progressing on the surface while submerged. The instrument comprises a tube fitted with crystal prisms in a special manner."

JANUARY 1902

"The curious properties of polonium, actinium and radium are for the most part opposed to all accepted mechanical theories, physical and chemical, for they appear to be spontaneous producers of light and electricity, in a word, of energy. Now, it cannot be admitted that a body can produce energy indefinitely, however small the production, without borrowing from external sources, and without losing from its mass, and yet this appears to be the case with the three new metals. It is difficult to imagine electric conductibility in the absence of every material particle; and as their rays are conductors, it may be supposed that there is an ultimate form of very attenuated matter, which these radio-active bodies may be able to emit indefinitely without losing noticeably from their mass. However it may be, the spontaneity of the radiation remains an enigma, a subject of profound astonishment. There is ground for believing that the discovery of these bodies marks a new stage in the grand history of science, and that it supports the hypothesis of the unity of matter, which has commanded the attention of philosophers for 2,500 years. Science is yet in its crudest manifestations, and our minds are scarcely trained to grasp the fundamental phenomena which incessant researches are gradually unveiling."

MARCH 1902

"Sufferers from nervous complaints may have reason to bless the memory of certain great apes who have cooperated unselfishly with some British scientists and surgeons in a series of privately conducted experiments to demonstrate new facts about the brain. The animals were anesthetized, and tiny openings made in their skulls. After the wound had healed entirely one electrode from an electric battery was fastened to the wrist of the chimpanzee in the form of a bracelet and the other electrode, in the form of a fine platinum point on a spring, was brought to touch the outer surface of the brain. Thus the areas that controlled the movement of the organs and limbs of the body became mapped out bit by bit. If a certain part of the cortex of the frontal lobe of the brain received the current, the ape thrust out his fingers; the current applied to another place made him thrust out his tongue. After the experiments had been concluded it was found possible to make a map of the brain as to its function. Two cases of injury to human brains which have since been treated according to the knowledge obtained from these experiments proved that the discoveries of the motor centers furnished fair working bases for treatment of the patient."

JANUARY 1903

"Since a negatively-charged body exposed in the atmosphere becomes radioactive, apparently indicating the presence of some radioactive substance in the atmosphere, it occurred to C. T. R. Wilson to test whether any of this radioactive substance is carried down in rain. He boiled down freshly fallen rain to dryness, and found a radioactive residue. The radioactivity rapidly disappears, falling over 50 per cent in the first hour."

FEBRUARY 1904

"Some further important evidence with regard to the suspected gradual conversion of radium or radium emanation into helium is furnished by Sir William Ramsay and F. Soddy. Radium emanation was condensed in a liquid-air tube, and the liquid air was then removed by the pump. The tube showed no trace of helium, but showed a new spectrum, probably that of the emanation itself. After standing for four days, the helium spectrum appeared, and the characteristic lines were observed identical in position with those of a helium tube thrown into the field of vision at the same time."

MARCH 1904

"In describing his successful experiments in powered flight, Wilbur Wright says, 'After the motor device was completed, two flights were made by my brother and two by myself on December 17 last. The first flight covered but a short distance. Upon each successive attempt, however, the distance was increased, until at the last trial the machine flew a distance of a little over a half mile through the air by actual measurement. We decided that the flight ended here, because the operator touched a slight hummock of sand by turning the rudder too far in attempting to go nearer to the surface. The experiments, how-ever, showed that the machine possessed sufficient power to remain suspended longer if desired. According to the time taken of each flight a speed varying from 30 to 35 miles an hour was attained in the air.'"

APRIL 1904

"To the many methods of purifying, modifying and preserving milk must now be added a process for *homogenizing* it so that it will keep almost indefinitely without change in its physical condition. The pro-cess has been perfected and patented by Gaulin of Paris, and is com-ing into use in Europe. It is designed to reduce all the fat globules in milk to a very minute size, by means of pressure and concussion."

AUGUST 1904

"It is gratifying to learn that science has at length discovered the real cause of 'caisson disease'—the terrible scourge which is the dread of engineers where submarine or tunneling operations have to be carried on under a pressure greatly exceeding that of the normal atmosphere. Profs. Hill and Macleod have shown that the various symptoms displayed by victims of caisson disease are produced by the effervescence of the blood in the small blood vessels consequent on the escape of the excess of air which exposure to pressure has forced into solution, and which subsequently effervesces like the gas in a freshly opened bottle of sparkling wine. This escape of air from the blood vessels obstructs the circulation in the parts nearest them, and the nature of the bad symptoms displayed depends on the position of the blood vessels in which most air happens to be absorbed at the time, and in which effervescence is most readily effected."

JUNE 1905

"It has generally been conceded that there is a demand on the part of dairymen, soda fountain proprietors, saloon-keepers, butchers, and grocerymen for something that would make them independent of the iceman, and this seems at last to have been accomplished. An electric refrigerator has been in operation for some time as an experiment in a Philadelphia grocery store. When the store is opened in the morning, the current is turned on and remains so during the day. Although the box is being constantly opened and closed, the temperature is maintained at 34 degrees. When the store is closed for the night the current is shut off, and the temperature remains almost constant all night. The iceless refrigerator is much the same in appearance as any large refrigerator. The motor and compressors are at one end, and the place usually occupied by the ice is given over to brine, which is the means of cooling the interior of the refrigerator. No expert knowledge of either electricity or refrigeration is required to operate one of these outfits. The switch controlling the electric current is the only part that must be manipulated. Where it is desired to have ice for use on the table, these machines will make it, in ten-pound pieces or in small cubes, while performing their ordinary functions."

JULY 1905

"For ascertaining the depth of the sea without the use of the sounding lead or other devices, the Norwegian engineer Berggraf has invented a unique method. He sends sound waves perpendicularly into the water, and measures the time they require to return to the surface after having been reflected from the bottom of the sea. The speed of sound in water being known, the length of the space passed through is immediately determined; one half of it is the depth of the water. A period of four seconds, for instance, between the departure and return of the sound corresponds to a depth of 2,400 meters."

JULY 1906

"The recent summer meeting of the American Association for the Advancement of Science at Cornell University was rendered memorable by the dedication of the largest and best-equipped physical laboratory in America. Prof. Wallace C. Sabine, spoke of neglected factors in determination of musical quality. When a complex tone is sounded, the fundamental tones do not die away so soon as the overtones, and it is found that the material of which the walls of an auditorium is constructed has a material effect in deadening the overtones, and thus changing the quality of the music or of the voice. Hence it seems that more attention should be paid to the material with which the walls are covered."

FEBRUARY 1893

A vessel which stands alone in its class, the first vessel of our new navy built solely to be used as a ram, was launched at the Bath Iron Works, Me., on February 4. It was christened Katahdin, after Mount Katahdin, the highest mountain in Maine.

The Katahdin is a twin screw armorplated vessel, built from the designs of Rear Admiral Daniel Ammen, and is based upon the personal experience of the admiral in the use of and defense against rams in our civil war, 1861-65.

Length over all, 251 feet; extreme breadth, 43 feet 5 inches; total depth from the base to the crown of deck amidships is 22 feet 10 inches; and normal draught of water is 15 feet, the corresponding displacement being 2,155 tons. The curved deck is armorplated throughout, the thickness of the armor tapering from 6 inches at the knuckle to 2 inches at the crown of the deck. Above this deck is a conning tower of 18-inch plate.

A continuous water-tight inner bottom two feet from the outer skin is carried nearly the whole length of the vessel and up to the armor shelf on each side. There are seventy-two water-tight compartments.

The ramhead is of cast steel extending back 11 feet, and it is supported by longitudinal braces in such a way that the force of the blow delivered by it is designed to be distributed through the vessel. The maximum estimated speed, at full power, is 17 knots, and the impact of the ram is designed to be equivalent to the blow of a hammer weighing over two thousand tons moving at this rate of speed—a blow which, if fairly delivered, would crash through the sides of any vessel afloat.

NOVEMBER 1906

"Some little discussion has taken place recently regarding the possibilities of 'seeing electrically.' Selenium when carefully prepared becomes an electrical conductor whose resistance is affected by incident light. If a selenium surface were divided into a large number of small squares each of these would represent the small squares of a half-tone. If, from each square, a wire be led out and an equal source of electromotive force be introduced into it, the current flowing will correspond to the intensity of light on the corresponding square and determine the amount of illumination which should fall upon the corresponding square of the receiving surface. The final picture would be built up of a large number of small squares and would resemble a half-tone reproduction. It is, of course, out of the question to carry a considerable distance any such numbers of wires as this arrangement would require. However, the impressions might be produced successively with sufficient speed requiring that two different mechanisms, one at the transmitting station and one at the receiving station, run in absolute synchronism. At the present time, we seem to be a long way from accomplishing this result."

MAY 1908

"For some months past the British military authorities have been experimenting with a new type of tractor for the haulage of heavy vehicles over rough and unstable ground. This machine, represents a new development in traction. Briefly, its object is to crawl over the ground, there being a series of feet disposed along the periphery of two heavy side-chains passing over fore and aft wheels. As this chain revolves, the feet are successively brought into contact with the ground, thereby impelling the machine forward or backward. Because of its peculiar movement, the soldiers at the Aldershot military center, where it is in operation, promptly christened it 'the caterpillar.'"

JUNE 1908

"Within a short time a Marconi wireless station will be established on the roof of the Bellevue-Stratford Hotel in Philadelphia, so that guests may communicate with their friends at sea. If the plant works successfully, a similar one may be put into operation on the roof of the Waldorf-Astoria in New York."

OCTOBER 1908

"Upon the new observations of Prof. Hale, made at Mount Wilson, Calif., on the double lines in sunspot spectra, Prof. Zeeman bases a theory that sunspots are strong magnetic fields. The source of light in a magnetic field emits two rays circularly polarized in opposite directions and parallel to the lines of magnetic force, according to Prof. Zeeman's experiments. The sunspot lines photographed by Prof. Hale are identical in character with these lines. To produce the actual phenomenon observed would require a current of about 5,000 amperes. The theory throws a great light upon meteorological and terrestrial magnetic phenomena, affording, as it does, some reason for the perturbations observed in the electric and magnetic equilibrium of our earth and its atmosphere."

FEBRUARY 1909

"Oxybenzyl-methylenglycol-anhydride is the chemical name of a coal-tar product which is being used as an insulator. However, it goes by a trade name of bakelite after the inventor, Dr. L. H. Baekeland. It is stronger than hard rubber, withstands a higher temperature, and is unaffected by most chemicals."

JUNE 1909

"Prince Henry, brother of the German Emperor, is the inventor of an automatic window washer. It is intended for the purpose of wiping off moisture from the glass wind-break of an automobile, so that the rider's vision may be clear at all times."

JULY 1909

"Pipe-line connections have been completed by which it is possible to pipe oil from Oklahoma wells to New York harbor. Oil has been started on the long journey of 1,500 miles. This is the longest pipe line in the world. It is not probable that much oil from the mid-continent district will be brought to the seaboard at present, and the completion of the line seems to be more in the nature of a provision for the future. Oklahoma has the most active oil field in the country at present, moreover its production is increasing, while that of Pennsylvania and West Virginia is decreasing. It may not be long before the western wells will be called upon to supply the seaboard and export demand."

SEPTEMBER 1909

"After Commander Peary's expedition of 1906, when he reached 87 deg. 6 min. N. lat., then the 'farthest north,' he determined to make one more effort to reach the pole. The *Roosevelt*, equipped by the Peary Arctic Club with all the material and scientific instruments which have been proved to be most essential in polar exploration by Commander Peary's 23 years of experience, left New York on July 6th, 1908. The ship proceeded via Etah in Greenland to Cape Sheridan in Grant Land and there the expedition passed the winter. The sledge expedition for the Pole left the *Roosevelt* in three divisions on February 15th, 21st and 22nd, the total of all divisions being seven whites and 59 Esquimaux, with 23 sledges drawn by 140 dogs. All the divisions assembled at Cape Columbia by February 27th. Open water delayed the entire party till March 11th, when the lead was sufficiently frozen over to be crossed. A few days later the ice parted exactly where the main party was encamped, nearly causing the loss of dogs and sledges, but after an exciting period dashing from one moving floe to another, better going was reached. After having crossed the 88th parallel Capt. Bartlett, the sturdy navigator of the *Roosevelt*, who has borne the brunt of the pioneering work, reluctantly turned back with the two Esquimaux of the last supporting party, the provisions carried being insufficient to last more than six men and 40 dogs for the week or more estimated to be required to reach the Pole and return. Peary then determined to try and reach the Pole in five forced marches allowing less than a day for each. An observation at noon on April 6th showed that they were only a little over three miles from the Pole, so the remaining distance was apparently covered before a rest was taken. The first 30 hours at the Pole were spent in making observations and taking pictures. A sounding was made through a crack in the ice five miles from the Pole, 1,500 fathoms of wire finding no bottom. Speed was urgent on the return journey and Peary was singularly fortunate in escaping open leads in the ice, which had delayed the return of the supporting parties. By continued rapid traveling, Cape Columbia was reached on the 23rd of April. On July 18th the ice was sufficiently open for the *Roosevelt* to be removed from her berth. She fought her way south and reached Indian Harbor September 5th to send the now historic telegram: 'Stars and Stripes nailed to North Pole.'"

OCTOBER 1909

"Halley's comet, for which astronomers have been watching so eagerly, has again been discovered. It was detected by Prof. Max Wolf on a photographic plate he had obtained by means of the 30-inch reflecting telescope at Heidelberg, Germany, on Saturday night, September 11th. One day later the observation was confirmed by

Heber D. Curtis, who succeeded in photographing the comet with the Crossley reflector at the Lick Observatory, Mount Hamilton, Calif. The location of the comet on the photographic plate agreed exactly with that indicated by Prof. Wolf."

JANUARY 1910

"Convincing evidence that the automobile of to-day is as far perfected as the materials of construction and mechanical ingenuity will allow is afforded by the fact that the cars shown in the two annual exhibitions this year exhibit no novelties of a radical character as compared with the cars of the preceding year. The tendency toward standardization is even more marked this year than last, and the freak car is conspicuous by its absence. For all cars the four cylinder, four-cycle engine, with variations in the valve arrangements, has become the standard type. Undoubtedly the present prosperity in the automobile industry is due largely to the fact that many people of moderate means, who have been waiting until a thoroughly serviceable car was placed on the market at low price, are now being accommodated. Several makers are offering a four-cylinder 20-horsepower car, having all the features of stylish design and certainty of control of the more costly designs, and for the low price of $750."

FEBRUARY 1910

"The infectious nature of poliomyelitis has been assumed rather than proved; it would now seem that complete demonstration of infectivity will presently be forthcoming. Early in 1909 two German experimenters, Landsteiner and Popper, successfully inoculated two monkeys with spinal cord taken from two fatal human cases of poliomyelitis. In both the monkeys lesions of the spinal cord were found similar to those in man. In September of 1909 Dr. Simon Flexner and his colleague, Dr. Paul A. Lewis, of the Rockefeller Institute in New York City, similarly inoculated monkeys with emulsions of human spinal cord and later with emulsions of the cords of monkeys that had developed paralysis after injection of the first emulsion. In one series, seven monkeys were each successively inoculated with the virus from the cord or cortex of its predecessor, the disease regularly resulting. Flexner and Lewis have found that the virus of infantile paralysis is of the same nature as that of smallpox; it belongs to the class of the minute and filterable viruses. There should be no reason in science why a vaccine or an immunizing agent against poliomyelitis should not in good time be forthcoming."

APRIL 1910

"Whereas biology was until recently chiefly a science of observation, it has now become in a high degree experimental. Gregor Mendel, 40 years ago in his cloister at Brünn by his careful experiments on the crossing of thousands of peas, succeeded in unveiling a law which has profoundly influenced ideas on heredity, not only in plants but in animals. We here have a definite arithmetical relation, which is susceptible to very exact study and confirmation. The method of Mendel, which we may call that of experimental evolution, is now of wide application, and there are laboratories which do nothing else but breed and cross under very exact control."

MAY 1910

"Now that aircraft have been entered as war vessels, inventors are beginning to cast about for some effective means of destroying them. Recently an aerial torpedo has been invented, which, by means of a hertzian-wave controlling system, may be directed from a distance without carrying any operator. This torpedo was exhibited at the London Hippodrome, where the inventor caused it to travel out over the audience, steering it wherever he chose by pressing buttons on a switchboard on the stage. The device may be equipped with explosives, to be dropped on an enemy. This was demonstrated by releasing flowers on the audience."

JUNE 1910

"Prof. Fritz Haber claims to have solved the problem of the direct synthesis of ammonia from its elements, nitrogen and hydrogen. If the process is as practical and economical as its inventor claims, its intro-duction will quickly cause a revolution in a comparatively new but already important branch of industry: the manufacture of artificial nitrates. Prof. Haber states that the combination of hydrogen and nitrogen is effected at a temperature of about 1,000 deg. F., and a pressure of 200 atmospheres. The presence of a catalyzer is required to accelerate the combination. For this purpose Prof. Haber employs ura-nium, but the rarity of this element appears incompatible with its employment on a commercial scale."

AUGUST 1910

"It is gratifying to learn that twice during the past three months the Bureau of Public Health of Philadelphia has been able to report that no deaths occurred from typhoid fever in a whole week. During the month of June there were but 13 deaths from this scourge as com-pared with 669 deaths during the corresponding month of 1906. Philadelphia has reason to be proud of the results gained through the adoption of filtered water, the deaths from typhoid having decreased steadily from 72.4 per 100,000 in 1906 to 21.2 per 100,000 in 1909. The indications are that the present year will show a still more strik-ing fall in the death rate of this disease."

CHAPTER FOUR

FEBRUARY 1911

"By means of the new submarine telephone cable from Dover to Cape Gris Nez, on the French coast, and suitable land lines, it will be possible to carry on a conversation from two ends of the wires in towns 850 miles apart, and it will be easy to speak from London to St. Petersburg. By introducing small self-induction or loading coils into each of the wires at spaces of about one mile apart, the defects of indistinctness and weakening of the sound noticeable in long cables of the old time have been overcome."

MARCH 1911

"The Norwegian navy has recently been strengthened by the acquisition of a new submersible, the *Kobben*. This vessel is of the *Germania* type, evolved and developed by Fried. Krupp A. G., the eminent German armament manufacturer and naval builder, which has now become the standard class of submarine in the Imperial German navy. In fact, the *Kobben* may be considered the latest development of the Krupp submersible, and is not only powerful but possesses many interesting features. The recent trials of the craft have aroused keen interest among European naval circles, owing to her striking seagoing qualities and general technical perfection. The Krupp firm have always considered the 'diving' boat to be the most efficient form of submarine. It may be mentioned that their contentions as to its all-round superiority are upheld by the German government's naval authorities, for there were 12 submersibles fitted for service on the the high seas in the Imperial navy at the end of 1910, which on account of their speed and endurance are suited to all the conditions of modern war."

APRIL 1911

"On February 10 the French Senate passed a bill that makes Greenwich time legal in France. When the law goes into effect French time will become nine minutes and 21 seconds slower than it is now. In order to avoid the expense of altering charts and sailing instructions, the law will not apply to French naval or other vessels, and it is not likely that any change will be made in the French almanacs. French railways are now run by a standard five minutes slower than Paris time, and the clocks inside stations are regulated by this standard, while the clocks on the outside of the stations give the correct Paris time. This confusing system will be abolished, and both the exterior and the interior clocks will be regulated by Greenwich time, by which the trains will be run."

JUNE 1911

"Svante Arrhenius has advanced an ingenious theory to account for the glacial periods that have marked several stages of geological history. According to the experiments of Langley, the carbon dioxide and the water vapor that the atmosphere contains are more opaque to the heat rays of great wave length emitted by the earth than to the waves of various lengths emanating from the sun. Arrhenius infers that any increase in the proportion of carbon dioxide and water vapor in the atmosphere will increase the protection of the earth against cooling and will consequently raise the temperature of its surface. The theory assumes that the earth's atmosphere was poor in carbon dioxide and water vapor during the glacial periods and rich in these gases during hot periods."

SEPTEMBER 1911

"The Otis Elevator Company has obtained a patent for an elevator that has a conveyer transporting between different levels in which the direction of movement is clockwise and contraclockwise. The elevator is in the form of a moving stairway, having ascending and descending series of steps that travel in spiral paths in opposite directions about a common center of curvature. The inventor is Charles Leeberger of New York City."

OCTOBER 1911

"One of the prominent automobiles for 1912 will be equipped with a generator and storage battery normally used for lighting the lamps and igniting the engine, but with the generator so arranged that it may also be used as a motor to 'turn over' the engine, thus obviating the necessity of cranking by hand. When the operator pushes the clutch pedal, a gear wheel on the electric motor will engage with teeth on the flywheel, and the motor will be operated by current from the storage battery, to turn the flywheel and start the engine. When the engine starts, the motor becomes a dynamo and generates current to be used for charging the storage battery and for ignition purposes."

MARCH 1912

"The first attempt that has ever been made to drop from an aeroplane in a parachute occurred at St. Louis on March 1. When flying with Anthony Janus in a biplane at a height of some 1,500 feet, Albert Berry climbed out of the seat and cut loose his parachute, which was suspended beneath the machine in a special case. He dropped suddenly for some 500 feet before the parachute opened, but it opened successfully and Berry reached the ground without hurt. Janus was able to control the machine when it was suddenly released of his companion's weight, and he too descended without mishap."

OCTOBER 1912

"A bacterial epidemic has within two years freed Yucatán of the locust swarms that periodically invaded the country. M. d'Hérelle, having been asked by the Argentine government to test the effects of the same microbe on another locust species that every year devastates large portions of the Paraná district, has reached surprisingly favorable results. Tests made on a large scale were quite as successful. The speed with which the malady was spread can be inferred from the fact that a few days after the first infection it occurred at a distance of 50 kilometers (31 miles) from the center of infection."

DECEMBER 1912

"The brilliant Russian physiologist Pawlow has for some years been conducting an exhaustive investigation by scientific laboratory methods of the reflex action of animals. Certain results of his latest studies are interestingly résuméed by Prof. Lüthje in the *Deutsche Revue*, from which we quote: 'Pawlow now no longer speaks of psychoreflexes but of conditioned and unconditioned reflexes. The latter are those that invariably occur when the appropriate stimulus finds a sensory path, as when food is put in the mouth and a flow of saliva follows. A conditioned reflex, on the other hand, is one that occurs only under certain given circumstances: if food is frequently shown to a dog and afterward given him to eat, after a certain number of experiments a flow of saliva will occur at the mere sight of the article (a "natural conditioned stimulus"). "Artificial conditioned stimuli" have the same effect. If a given musical note is repeatedly sounded at the same time that a given article of food is offered to a dog, after a certain lapse of time the mere sounding of the note will produce a corresponding flow of saliva. Similarly other external conditioned stimuli (optical, thermal, etc.) can be formed, if the same stimulus is repeated a number of times synchronously with an unconditioned stimulus, such as the taking of food.'"

MARCH 1913

"On Monday of last week the Edison Kmetophone was exhibited for the first time in public on the stages of four prominent vaudeville theaters in New York City. In the first film a man made a speech explaining the perfecting of the talking picture by Edison in the obtaining of absolute synchronism between the pictures and the sounds. Then a

Pianist played, a bugler sounded the reveille, a young lady sang and some dogs appeared and barked. A sound picture showed a minstrel enter tainment in which the various member of the troupe performed as naturally as in real life, although there was no mistaking the fact that the talk and music were produced by a phonograph. The intensification of sound necessary in a theater apparently gives to it more of that metallic quality of which it has been successfully deprived in the smaller drawing-room machines. Nevertheless, it is no more disagreeable than the voices of many of the actors who appear on the vaudeville stage. The present talking pictures last about five minutes each."

MAY 1914

"Two French engineers, A. Papin and D. Pouilly, are responsible for a new departure in aeronautics—a departure that has not yet departed from the solid earth. They explain that in designing their machine they had before them mainly three purposes: to provide a device which could rise from the ground without preliminary 'run' and which could similarly alight on a selected spot; to furnish a machine that could at will either advance through the air or be held stationary; to provide for a slow descent in case of failure of the motor. The new 'flying machine' constitutes a huge single-blade screw propeller with a spread of 178 feet. The blade and the central body are hollow and through them a stream of air is blown by a fan, which is driven by a motor. The motor is an 80-horse-power, nine-cylinder gasoline engine and is manipulated by compressed-air controls from the car, which is mounted on ball-bearings and does not participate in the rotation of the rest of the machine. The blast of air escaping from the curved tip of the propeller causes it to revolve by impulse reaction."

JULY 1914

"A. H. Pfund of Johns Hopkins University, describes some preliminary tests he has made of a new apparatus for measuring the light of a star. The work was done at the Allegheny Observatory with the Keeler 30-

inch reflector. In the focus of the telescope was placed either of two small blackened disks which formed the junction of a thermo-circuit. The wires used for the thermo-element were enclosed in an evacuated capsule closed at one end by a plate of fluorite and substituted for the eye-piece of the telescope. The thermo-current was measured by a moving-coil galvanometer. The sensitiveness of the arrangement was such that a candle at a distance of eight miles would give a deflection of one millimeter. The deflections obtained from celestial objects were: Vega, 7.5 millimeters; Jupiter, 3.0; Altair, 2.0. The author hopes, by using a more sensitive galvanometer and other materials for his thermo-elements, to increase the sensitiveness considerably and in this way to open up a new field of astrophysical research."

AUGUST 1914

"The Sperry gyroscope, in its application to the flying machine, is another example of the remarkable speed of the development of some modern epochal inventions. Two years ago Mr. Elmer A. Sperry fitted a Curtiss aeroplane with his device and experiments in stabilizing were then undertaken. Full details have now been received of the signal triumph of this wonderful product of American ingenuity in France, where, at a safety the French war department was won by the aeroplane fitted with the Sperry stabilizer. Lawrence D. Sperry, son of the inventor, piloted the winning machine, assisted by a mechanician. The control of the machine was perfect, young Sperry standing up during the flight with arms folded. While the mechanician climbed to the end of the lower plane and back."

APRIL 1915

"Until recently it had been assumed that a given chemical element must always possess the same atomic weight. The inclusion of the radioactive elements in the periodic system, shows that we must assume the existence of elements which vary as much as eight units in atomic weight, with corresponding variations in their radioactive properties but without any change in their chemical behavior. This conception was based on indirectly proved, or inferred, properties of the short-lived radioactive elements. Hence it was important to prove by direct experiment that two elements that appear chemically identical may have different atomic weights. The atomic weight of the lead formed in uranium ores in the course of millions of years as the final product of the disintegration of uranium probably differs from the atomic weight of lead extracted from common lead ores. This conclusion has now been confirmed experimentally by very careful determinations of the atomic weights of specimens of lead of diverse origins. The research was carried at Harvard University, which is celebrated for its accurate methods of determining atomic weights. The atomic weight of lead obtained from uranium ores was found to be 206.6, whereas ordinary lead gave the distinctly different value of 207.1."

MAY 1915

"Einstein's incessant efforts to improve his theory of gravitation in its first form have resulted in the admirable theory which he has recently published. The improved theory leads to the conclusion that gravitation is propagated with the velocity with which light moves in the absence of a gravitational field. Einstein has pointed out some results of the theory that may perhaps make it possible to observe almost directly the variation of luminous velocity in a field of gravitation. First: A luminous pencil should be curved by the influence of weight. Einstein calculates that a ray of light coming from a star and grazing the sun's surface would be bent inward by .83 second, increasing by that amount the apparent angular distance of the star from the sun's limb. This effect might possibly be observed in a total solar eclipse. Second: If light coming from two sources of different heights is examined with the same spectroscope, the spectral lines of the higher source should be a little nearer the violet than the corresponding lines of the lower source. For two similar molecules, situated respectively on the sun's surface and at the earth's distance from the sun, the difference is about a hundredth of an angstrom unit. Hence the Fraunhofer lines of the solar spectrum should be nearer the red by this amount than the corresponding lines of the lower terrestrial source.

Displacements of this order of magnitude have actually been observed. They have been attributed to effects of pressure and movement, but they may be due to the cause indicated by Einstein."

OCTOBER 1915

"The possibility of devising an electrical machine for solving numerical equations to any degree has recently been suggested by a French author. Essentially the machine will consist of a collection of various electromagnetic machines connected in cascade, the armature circuit of one machine being used in the excitation circuit of the next and so on. It has been demonstrated by a commentator on the suggestion that by the connection of transformers in cascade it would be possible to solve not only algebraic equations but linear differential equations with constant coefficients as well."

FEBRUARY 1916

"The late Dr. Aksel Steen, director of the Norwegian meteorological service, had charge of working up the magnetic observations made by Amundsen on his northwest passage of some years ago. *Terrestrial Magnetism* publishes a letter written by Dr. Steen shortly before his death last May, stating that two or three years more would be required to complete the work. The writer declares that it will be impossible to give a definite position for the north magnetic pole, because in his opinion this pole 'is not a fixed point attached to a certain geographical latitude and longitude but must be defined as that point on the surface of the earth where the horizontal intensity at the moment is zero.' The discussion of Amundsen's observations will probably show that the pole has a mean daily and yearly periodic motion, together with more or less irregular displacements."

OCTOBER 1916

"The most novel, if not the most spectacular, feature of the recent successful offensive by the French and British armies on the Somme was the presence of several armed and armored tractors of the caterpillar type, which, if we may judge from the press reports, proved wonderfully effective in following up the heavy gun attack, riding down or cleaning out machine-gun emplacements, enfilading trenches and otherwise preparing the way for the rush of infantry attack. What part these machines are destined to play in the later stages of the war is a matter of pure speculation. The British speak of them as a great success; Berlin naturally describes them as being a complete failure—unwieldy, slow and liable to break down. If they are successful, Germany is certain to come back with something of the same kind; and if so, we may see squadrons of these mechanical armadillos maneuvering against each other in the open field."

NOVEMBER 1916

"The future use of the selenium cell in astronomy is discussed by M. Fournie d'Albe in a recent contribution to *Scientia*. The author finds that a selenium cell having a surface of 100 square centimeters would theoretically be capable of registering the light of a 28th magnitude star, which is, of course, far fainter than any visible in the larger telescopes. For such observations, however, an exposure of several days would be required, and this would be difficult to accomplish. On the other hand, with an exposure of only one second such a cell is much more sensitive than the eye. The author believes that we can virtually increase the diameter of our greatest object glasses tenfold by substituting a selenium cell for the human eye at the eyepiece of the instrument. Thus it should be possible to detect stars about five magnitudes fainter than any now observable."

MARCH 1917

"A steel that does not stain or tarnish is one of the latest new materials and will be welcomed by the housewife as a real boon. It is called 'stainless steel' and from it table cutlery is being made that not only takes a beautiful polish but also preserves this appearance under all circumstances. The new chromium-alloy steel was discovered in England but is now being made in the United States and sold as table cutlery. Its possibilities, however, are by no means limited to cutlery. One can readily imagine to what countless uses a stainless and rustless metal can be put."

DECEMBER 1917

"It is announced from Mount Wilson that a new star of about the 14th magnitude has appeared in the spiral nebula N.G.C. 6946. In connection with this discovery, Dr. H. D. Curtis points out that this is the sixth time the appearance of a nova has been reported in spiral nebula. The number of these cases indicates that the new stars in question were actually in the nebula, and not merely in a line with it as seen from our system."

JANUARY 1919

"The development of a hydrogen substitute that would be non-inflammable has always been considered by airship advocates as the most important progress that remained to be realized in aerostatics to make the airship really safe. Heretofore the difficulties encountered in this endeavor have appeared as an unsurmountable stumbling block. Today the great problem is solved, for American enterprise, engineering skill and ingenuity have succeeded in achieving an extraordinary *tour de force* by developing apparatus for the production of helium in large quantities and at a comparatively low cost. Helium, an inert, non-inflammable gas, the second lightest known (the lightest being hydrogen), is relatively abundant, but the operation of separating it has involved such a great expense—from $1,500 to $6,000 per cubic foot—that its use as a hydrogen substitute was never seriously considered until the war. When it is considered that by next spring helium will be produced in this country on an industrial basis and at a cost of approximately 10 cents per cubic foot, the magnitude of the achievement will be fully realized."

MAY 1919

"An account of the discovery by Major H. Graeme Gibson, Major Bowman and Captain Connors of the Allied army medical services of what is stated to be very probably the causative germ of influenza appeared lately in the London *Times*. The germ belongs to the order of filter-passers and is grown by the Noguchi method. The discovery cost Major Gibson his life, as he fell a victim to the very virulent strains of the germ with which he was experimenting."

DECEMBER 1919

"Dr. Einstein tells us that when velocities are attained which have but just now come within the range of our close investigation, extraordinary things happen—things quite irreconcilable with our present concepts of time and space and mass and dimension. We are tempted to laugh at him, to tell him that the phenomena he suggests are absurd because they contradict these concepts. Nothing could be more rash. When we consider the results which follow from physical velocities comparable with that of light, we must confess that here are conditions which have never before been carefully investigated. We must be quite as well prepared to have these conditions reveal some epoch-making fact as was Galileo when he turned the first telescope upon the skies. And if this fact requires that we discard present ideas of time and space and mass and dimension, we must be prepared to do so quite as thoroughly as our medieval fathers had to discard their notions of celestial 'perfection,' which demanded that there be but seven major heavenly bodies and that everything center about the earth as a common universal hub."

JANUARY 1920

"Papers by Prof. E. B. Wilson and Dr. Leigh Page read before the autumn meeting of the National Academy of Sciences led to a discussion of the announcement recently made in England that the results of the British eclipse expeditions to Brazil and West Africa had confirmed the existence of a deflection of the light of the stars passing by the sun of the same magnitude as that predicted by Einstein's generalized theory of relativity. The discussion developed the point of view, especially among the experimental physicists present, that it is not necessary to suppose that the deviation of the light observed is really a proof of the soundness of the theory of relativity. Several other causes, such as diffraction and refraction, might be operative to produce the small bending of the star rays seen. The impression prevailed that, so far as observations of any kind have yet gone, no one is compelled to adopt the difficult and obscure theory of relativity in place of other explanations of phenomena unless he has the type of mind that prefers it."

FEBRUARY 1920

"The public was startled recently by newspaper announcements that a rocket had been invented which would carry as far as the moon. Sensational as this statement appeared to be, it was nevertheless issued by the Smithsonian Institution and was based on the work of Dr. Robert H. Goddard of Clark University, who has been conducting a long series of experiments on existing forms of rockets. He has developed a method of increasing the efficiency of this type of projectile to such an extent that it will be possible to propel a rocket beyond the influence of the earth."

MAY 1920

"Much interest has been aroused by a hypothesis that is usually called the 'electron theory.' One of the latest versions of this theory has been elaborated by Dr. Irving Langmuir. The Langmuir theory can be stated very briefly as follows: The atom of each element is composed of a nucleus and electrons. The number of electrons is in every case given by the atomic number. The nucleus is surrounded by electrons, which are arranged about it in concentric spherical shells. Each electron possesses a negative electric charge. The nucleus contains as many positive electric charges as there are electrons. The entire atom is therefore electrically neutral. With the exception of hydrogen every element has the following arrangement. Its nucleus is surrounded in the first shell by two and only two electrons, the other electrons going into the outer shells. The outer layers of the electrons that go to make up an atom tend to form groups of eight; these groups, represented diagrammatically as cubes When two octets are adjacent to each other in the architecture of a pair of atoms, the theory indicates that they may hold in common one or two *pairs* of electrons but never a *single* electron. The property which causes octets to share pairs of electrons is called covalence. Atoms unite because each one which has an insufficient number of electrons in its outer shell to form a complete octet endeavors to complete its octet by sharing pairs with the finished or unfinished octets of other atoms."

JANUARY 1921

"Mr. Lee De Forest, reviewing the development of the audion in an extensive paper read before the Franklin Institute, brings out some very interesting applications of this apparatus and forecasts the use of it in many new directions. Most striking is the suggestion of producing music electrically. To quote the author: 'The uniform generation of electrical oscillations in a circuit by means of an audion is one of the most fascinating of its applications. If these are of radio frequency there is no sensible manifestation of their presence, but if of audio frequency the telephone receiver or loud-speaker reproducer may be made to give forth sounds from the highest pitch or volume to the softest and most soothing tones. Such wide range and variety of tone can be produced from suitably designed singing circuits that a few years ago I prophesied that at some future time a musical instrument, involving audions instead of strings or pipes, and batteries instead of air, would be created by the musicians' skill.'"

FEBRUARY 1921

"Public attention has been widely and deservedly attracted during the past month to a notable achievement of observational astronomy—no less an advance than the actual measurement of the diameter of a fixed star. With a high magnifying power all stars appear as minute disks, but the size of these disks depends not on the star but on the telescope. They are mere optical effects, inevitably produced by the influence upon the waves of light of the circular aperture through which the light enters the telescope. The greater the aperture of the telescope is, the smaller will be this 'diffraction pattern'; but even with the largest instruments it remains of such a size as to conceal from us those fine details which often we would most desire to observe—such as the existence of very close double stars and in particular the actual disks of the stars themselves. Professor Michelson, by a brilliant application of the interferometer, has added the solution of this problem to the scientific advances which this apparatus has permitted. The apparatus which has just been put into use at Mount Wilson, as an attachment to the 100-inch telescope, permits of a maximum separation of 20 feet between the interferometer mirrors. With this apparatus it has been determined that the diameter of the star Betelgeuse is probably fully three times as great as the earth's distance from the sun—in round numbers about 300,000,000 miles, fully as big as the whole orbit of Mars."

MAY 1921

"The problem of the world's supply of energy is the subject of an interesting discussion published recently by Dr. Arrhenius. It is pointed out that the early exhaustion of our fossil fuels will require the use of such other sources of power as water, wind and sun. The estimated life of the coal fields is put down at 1,500 years, and he believes it to be clear that we must soon ration our coal and substitute as far as possible other sources of energy. In view of the greatly increased use of petroleum. It is considered doubtful that mineral oils will constitute an adequate auxiliary supply. Dr. Arrhenius calculates that the continual increase of carbon dioxide in the atmosphere from the burning of coal will give the whole world a more uniform and warmer climate."

JUNE 1921

"Medieval theologians have been ridiculed because they debated how many angels could stand on the point of a pin. Prof. R. A. Millikan of the University of Chicago gives science's answer to a modern problem that is more or less comparable to this one when he isolates and measures an electron; he has recently been catching individual atoms and counting the number of electrons which each one has lost when an alpha particle from radium shoots through it. Science for some time has divided the 'indivisible' atom into its constituent parts and identified certain of these as electrons, but Prof. Millikan is the first to catch and exactly measure the charge carried by each one of these particles. Moreover, he can count the exact number of them which he has caught in a minute oil-drop, with quite as much certainty as he can enumerate his fingers and toes!"

NOVEMBER 1921

"Is air travel already wearying us and becoming like an old song? At the Chicago Pageant of Progress the 11-passenger hydroplane 'Santa Maria' carried a projector and an operator to beguile the tedium of flying with motion pictures while the hydroplane and the audience were hurtling through the air at 80 miles per hour."

JANUARY 1922

"Most unusual effects can be obtained by taking motion pictures at a high rate of speed and projecting them at the standard speed of 16 pictures, or frames per second. There are various special cameras for this kind of work, but until now their construction has been considered the deepest secret. A new high-speed camera for motion-picture work, designed to overcome the objectionable features of the make shift devices heretofore use, has been invented by a Pacific Coast man. The chief characteristic of the new camera is a straight pull on the film, with a camera capable of taking 14 feet to film (224 separate frames) per second. It will be readily appreciated that there is a terrific strain on the film to say nothing of similar destructive influences on the camera itself. With the film magazines mounted at the back, the new camera takes the film in a direct line back to the take-up magazine, thus providing an instrument that has only one turn for the film. With a turning mechanism that is geared very high, each turn of the handle exposes seven feet of film. The handle is turned at the same rate as with an ordinary camera: two turns per second. Hence to the operator there is no difference in manipulation. Of course when such a film is projected at the ordinary rate of speed. Objects move very slowly, making the instruments of incalculable value in scientific industrial work."

FEBRUARY 1922

"The widespread introduction of motor vehicles has done much for highway transportation, so that today it is a powerful rival of the railroads of this country. But with the ever increasing use of motor vehicles the highways of the nation have become more and more taxed until now there is an overcrowded condition on many of our highway systems. This condition is really acute in large centers of population, such as New York and its surrounding cities and suburban districts. It goes without saying that such an increase in street traffic in large cities has called for rigid and effective traffic regulation. In the case of New York's Fifth Avenue a signal system of traffic regulation has been evolved. Signal towers have been erected for the control of traffic on Fifth Avenue and on cross streets in the most congested areas. These towers make use of signaling lamps. The system has been successful beyond a doubt and has accomplished much in the way of eliminating annoying and costly delay. Even on remote rural roads traffic has become exceedingly heavy. It seems that this condition could be relieved if the state highway departments widened these roadways so as to make it possible for four lines of vehicles to be operated over them at one time."

MARCH 1922

"Niagara marks the point where nature has found an outlet for the waters of Lake Erie, but man has found it convenient to order the matter otherwise. Niagara is not to become dry, but with the completion in late December of the Queenston-Chippawa power canal much of the water that has taken a 162-foot plunge will go over the cliff at a point where the head available for power generation will be 305 feet. Dredges have for nearly four years been cutting a channel through earth and rock in a wide swing around the falls. They have taken out 13,000,000 cubic yards of earth and 4,000,000 of rock. The hydroelectric plant's full power development in contemplation amounts to 500,000 horsepower. The turbines and generators are larger than any in use elsewhere."

JUNE 1922

"Recognition is being given these days to the invisible portion of the spectrum, which is known to exceed greatly in range the portion to which the human eye is sensitive. Certain animalcules that will survive exposure to violet light for four or five hours are killed by the ultraviolet within 15 seconds. The effect of these rays on the human skin, eye and so on is different only in degree and in no sense in kind. It therefore becomes an object to screen off these rays from ordinary light, if it can be done."

JULY 1922

"Recent investigations in the field of proteins and nutrition conducted by the Bureau of Chemistry of the United States Department of Agriculture have shown that different proteins vary widely in their nutritive value. A diet may furnish a sufficient amount of protein, fat, carbohydrates, salts and vitamins and yet fail to promote growth or sustain well-being unless the quality of protein is nutritionally adequate. Certain of the amino acids, of which about 19 have been found in proteins, are absolutely essential for growth and maintenance among which are lysine, cystine and tryptophane. A protein that is deficient in lysine or cystine, even though it contains all of the other amino acids, will fail from a nutritive standpoint. The deficiency, however, when properly understood, can be supplemented by the addition of other foods to the diet that contain adequate amounts of these essential amino acids."

JANUARY, 1923

"After a lapse of several years there is renewed interest in talking pictures—motion pictures accompanied by more or less realistic sounds.

Obviously the pictures and the sounds must keep in perfect step throughout; otherwise the results are sometimes ludicrous and at all times unconvincing. All the earlier attempts made use of cylinder or disktype phonographic records, but these methods leave much to be desired. It is pretty well admitted today that any successful system of talking motion pictures must combine the sound record and pictures on a single film; any other plan only leads to trouble. What appears to be the most promising method is to record the sounds photographically at the same time and on the same film as the corresponding motion picture. The film moves intermittently—frame by frame— through the motion-picture mechanism, and continuously before and after while it is moving steadily the sound record is made. For exhibition purposes the film goes through the projector mechanism frame by frame and steadily through the sound-reproducing device. A beam of light is passed through the sound record, and the varying degree of shadow, falling on a light-sensitive cell, causes a current of fluctuating strength to pass through a circuit that includes loud-speaking devices."

APRIL 1923

"We have just heard of the latest improved method of causing an airplane to take off and alight at low speed, while at the same time making it possible for the airplane to maintain the maximum speed in flight. The manner in which this extreme difference in speed is accomplished is by varying the surface of the wings by withdrawing the rear part of the wings from the front part to increase the surface, and sliding the rear part back into the front part, into which it telescopes, when the airplane is in flight and it is desired to attain the maximum speed. The altering of the surface of the wings is accomplished by the aviator turning a crank within the body, which he can do very readily. This new improvement in airplane construction is the invention of a Frenchman, M. Bille, and it seems to be a far neater and better solution than that of the variable angle of incidence devised by Hollander Lanzius, or even that of the slotted wing, invented independently by Mr. Handley Page and by Herr Lachmann."

MAY 1923

"Professor J. F. McClendon of the University of Minnesota recently discussed the world's supply of iodine in relation to the prevention of goitre. Practically all of the iodine of the earth's surface is in the sea, which contains about 60 billion metric tons of iodine in the form of soluble inorganic salts. Judging by the prevalence of goitre, there is often a deficiency of iodine in our food and drink. Omitting the details of the local distribution of goitre, there is a wide goitre belt across the nation that includes the mountainous and glaciated regions. Since the run-off from these regions has carried away so much of the soluble material, it seems likely that the goitre belt is a low-iodine belt. Since the sea contains the bulk of the iodine supply, the transfer of iodine from the sea to our food or drink should be increased. Salt might be made an important source of iodine in our dietary scheme. Salt could easily be prepared from sea water with the retention of the iodine compounds and at a cost not exceeding that of present day table salt."

JUNE 1923

"In a lecture delivered before the Geological Society of London, Prof. Arthur S. Eddington says, regarding the question of the evolution of the earth-moon system: 'After the evolution of the solar system, we naturally turn to consider the evolution of the earth-moon system. My impression is that nothing in recent progress suggests any doubt that the beautiful theory of Sir George Darwin is correct. The main features are that the moon at one time formed part of the earth and broke away. At that time the rotation period of the earth was between three and four hours. That period has since lengthened to 24 hours, owing to frictional dissipation of energy by lunar and solar tides, and the back-reaction of the lunar tides on the moon has caused the moon to recede to its present considerable distance. Modern research has enabled us to calculate the magnitude of this tidal friction at the present time, and to estimate the date of the moon's birth to some 1,000 million years ago.'"

SEPTEMBER 1923

"Rural free delivery is not the only means of extending the scope of Uncle Sam's service to communities off the main line of communication. The latest idea for making the mails more useful involves the mounting, on interurban trolleys, of ordinary mail boxes. At any point along the line letters may be posted in these boxes, to be removed by the postman when the car reaches a good-sized town. This is a long step forward from the condition where people strung out through the country are dependent on the once-daily passing of the mail carrier for their contact with the outer world."

OCTOBER 1923

"About a year ago a group of scientists, in particular Drs. F. G. Banting and C. H. Best, working in the physiological laboratories of the University of Toronto, announced they had discovered a preparation that possessed the marvelous property of lowering the sugar content of the blood of dogs, when it was injected into them by means of a hypodermic syringe. This discovery was epoch-making in the history of medicine. The extract was made from the pancreas of animals, particularly the dog, the rabbit and the unborn calf. The process of making the extract has been applied on a commercial scale, and in this country there is at least one drug house that is ready to supply insulin in regular quantities. Insulin is not a permanent cure of diabetes; patients must keep on taking the drug constantly."

NOVEMBER 1923

"We have gradually learned that electricity exists in two forms, the negative form, which is called an electron, and the positive form, which is now beginning to be called a proton. There is no other kind of electricity so far as we know. The material universe seems to be built of these two elements. Both the electron and the proton are exceedingly small, very much smaller than an atom of matter. Both probably have weight, though one is much heavier than the other. The proton weighs as much as 1,830 electrons. But it is not appreciably bigger, and some even think that it may be smaller than an electron. The fact is, we do not know much about it, except that it is the unit of positive electricity, just as an electron is the unit of negative electricity. Whether the proton is an ultimate unit, or whether it can be resolved into a close-packed assemblage of simpler ingredients, which would account for its remarkable weight or massiveness, remains for future discovery."

FEBRUARY 1924

"The recent discovery of hafnium, the latest addition to the list of chemical elements, was the result of investigations based on the latest and most advanced conceptions of atomic structure. The list of elements arranged in the order of their atomic number showed a break after No. 71, the element lutecium. No. 72 was lacking. D. Coster and George de Hevesy, working in Copenhagen, deduced that the unknown element would probably show great resemblance to element No. 40, zirconium, to which, according to the theory of Niels Bohr, it must be closely related. The two investigators examined the X-ray spectra of all zirconium minerals and in each case found, in addition to the characteristic lines of the known element, lines of another, unknown element in the position where the lines of element No. 72 should be. The two scientists succeeded in separating the new element from the zirconium and named it in honor of Copenhagen (called *Hafnia* in its Latin form)."

MARCH 1924

"The fish filet has brought a revolution in marketing. It began experimentally in Boston about two years ago, when haddock filets were wrapped in parchment paper and shipped without ice to New England towns during the cold weather. Summer brought the shipment to an end, for a very large part of the difference between the seaboard and inland prices of fish is freight not on food but on ice and offal. One fish dealer's faith in the idea was so strong, however, that he developed a compact box in which a tin container holding the wrapped filet could be shipped with a moderate quantity of ice. Success was immediate. Inland grocers and butchers found that people who knew ocean fish only by reputation became customers for these fish filets."

NOVEMBER 1924

"Instead of futilely berating a sentimental but callous world for denuding our forests of young trees to supply the great annual Christmas-tree market, the Northeastern Forest Experiment Station is trying out the possibility of growing trees purposely to supply that market. In a plantation of two-year-old Scotch pine that the station has just initiated for experimental purposes at Mount Toby, Mass., Norway spruce seedlings have been planted at the centers of the squares formed by the six-foot-by-six foot spacing of the young pines. Here it will be possible to test the feasibility of raising Christmas-tree stock in mixture with a pine plantation and of making the Christmas tree an industry."

JANUARY 1925

"What happens when we get out of breath or become exhausted? Dr. A. V. Hill, professor of physiology in University College London recently set out to answer this question. He discovered that there is a very close relation between the fatigue caused by exercise and the production of a certain chemical called lactic acid in the muscles that are being used. Dr. Hill has shown that fatigue is always accompanied by an increase in the amount of lactic acid in the muscles, whereas recovery from fatigue is always marked by a decrease in lactic acid."

FEBRUARY 1925

"Sir Ernest Rutherford recently presented results he has obtained in studying the structure of the atom. Sir Ernest showed how he has been able to smash up the atoms of many elements by bombarding them with powerful particles projected from radium atoms as these break up spontaneously. The projectiles are known as alpha particles and weigh about four times as much as an atom of hydrogen. They carry a double charge of positive electricity, but whenever they can steal two electrons from the atoms that they drive through in their swift flight they settle down into inert atoms of helium gas. By means of a new and more delicate apparatus Professor Rutherford has succeeded in disintegrating the atoms of many of the natural elements, namely nitrogen, aluminum, sodium, potassium, boron, phosphorus, fluorine, magnesium, silicon, sulphur, chlorine and argon. Some of the elements, notably oxygen, have so far stubbornly resisted his best efforts. This exceptional stability may account for the fact that oxygen forms half of the substance of the rocks, air and water taken together. When the elements are arranged in the order of their atomic weights, it is found that the even-numbered elements form 86 per cent of the earth's crust; oxygen is one of the even-numbered elements."

APRIL 1925

"A Chicago man has invested $200,000, in a laundry. It is not an ordinary laundry but one for motor cars. The automobile is first pulled through a 'shower' on an endless belt, then rubbed down by hand with chamois cloths."

SEPTEMBER 1925

"C. Francis Jenkins, radio photographic experimenter of Washington, has succeeded in demonstrating apparatus by which moving objects, including a Dutch windmill and motion picture film, were sent by radio for five miles and reproduced on a miniature screen, ten by eight inches. The transmitter was set up at station NOF, near Anacostia, D.C., and the receiver in Jenkins' laboratory in Washington. He predicts that the process will be perfected so that scenes at baseball games and prize fights can be broadcast over long distances."

DECEMBER 1925

"During the past few months the leading scientific journals have contained several interesting articles concerning the remarkable similarity in effect on the human organism between natural light and ultra-violet light, as well as certain foods containing large amounts of vitamins. The evidence so far indicates that we are probably on the eve of the conquest of the deficiency disease known as rickets. Rickets seems to be primarily due not so much to a deficiency of the bone-forming calcium and phosphorus in the diet as to the inability of the body to make use of what is provided. It is now known that ultra-violet light has the effect of enabling the body to make use of the available bone-forming elements. What actually takes place when the radiation strikes our bodies is not wholly known as yet. Some scientists believe the light activates substances at the surface by a photochemical effect, like that which enables us to take photographs. By means of the blood those substances are later carried inward to the bones, enabling the bones to accumulate the necessary stiffening substances."

MARCH 1926

"Rays that reach the earth from some unknown outside source have proved to be far more penetrating than X rays. The tests made at Muir Lake in California are described by Professor R. A. Millikan: 'At the surface of the lake our electroscopes showed a rate of discharge corresponding to 12.4 ions per centimeter per second. As we sank the electroscope to greater and greater depths this rate of discharge decreased continuously down to a depth of 50 feet, after which it became constant at a value of 3.4 ions per cubic centimeter per second. This meant that we had obtained much more unambiguous and thoroughly quantitative evidence than had before been available for the existence of a radiation so hard that, if it came from outside the atmosphere, it required 50 plus 23, or 73, feet of water to absorb it entirely. The 23 feet is the water equivalent of the atmosphere above Muir Lake. Since at all the altitudes at which we have now experimented we find no measurable variation in the intensity of these rays at any time between midday and midnight, we conclude that they shoot through space equally in all directions.'"

APRIL 1926

"For perhaps 100 years boats have been propelled in shallow waters by pumping water to the top of a tank and then allowing it to escape at a fair rate of speed from the bottom of the tank in a narrow jet. The reaction of the escaping jet was sufficient to drive the boat forward, even though the efficiency of the mechanism was poor. Is it possible that the principle of jet propulsion can be applied to the airplane? The French Service Technique de l'Aeronautique is testing out an airplane power plant invented by H. F. Melot that is based on the principle of jet propulsion, the energy of the gas being converted directly into propulsive effort without the mediation of connecting rods, crank-shaft and screw propeller. American investigations on the subject have put in doubt the possibility of obtaining any great efficiencies, but for exceedingly high-speed planes the simplicity of the apparatus and the elimination of the propeller lead us to think that further developments are worth watching."

JULY 1926

"In a paper read before the Physiological Society in England, E. D. Adrian of the University of Cambridge suggests the possibility of eavesdropping on the human brain by means of radio instruments, in which vacuum-tube amplifiers play an important part. Dr. Adrian said he believed that within the next few years it should be possible to read the main types of brain messages passing down from the brain via the nerves to the muscles. The passing of the messages down the nerves seems to cause an electrical disturbance, and Dr. Adrian's apparatus records on a rapidly moving photographic plate the impulses along a single fiber. He 'decodes' the nerve impulse by seg-regating a single fiber of the system. He points out that the sense organs in the skin that register temperature, touch and pain are too close together for easy segregation but that the fibers in the muscles are farther apart and can be used as a link between the brain and the radio recorder."

OCTOBER 1926

"We have all seen the automatic toasting machine in restaurants and in the sandwich shops that are springing up all over the country. These toasters are too large for home use, but now we have a device that can be placed on the dining-room table. This little brother of the restaurant toaster turns out the same brand of golden-brown toast with no burning. You simply drop the bread into the oven slot and depress the two levers. When the toast is done, up it comes and the current is automatically cut off."

NOVEMBER 1926

"One of the first hospitals in this country to depart from the customary white operating room is St. Luke's Hospital in San Francisco. When the new hospital was constructed, the floor and wainscot of one of its operating rooms were finished in green tile. This was the idea of the chief surgeon, who called attention to the fact that the particular shade of green is the complementary color to 'blood' red. In this fact is to be found the reason for departing from the conventional white as an environment for the surgeon. Its dazzling brightness is blinding to the surgeon's eyes when he raises them from an operating field where the predominating color is red. Complementary colors afford the greatest eye relief and dark green proves to be the most efficient in combatting the color fatigue to which the surgeon is subjected. In addition to lessening color fatigue complementary colors also intensify their opposites. Thus the operating field stands out more clearly in the surgeon's vision after he gazes for a moment at green."

MARCH 1927

"Anti-knock fuels will form an item of great importance to motorists during the approaching summer season. It is well known that the knock encountered in an automobile engine working under load or badly carbonized reduces its efficiency. Because of this the use of antiknock fuels in an ordinary motor increases its efficiency as measured in the number of miles per gallon obtained from the fuel. The two types of antiknock fuels that now enjoy the widest use are those based upon 'cracked' gasoline and those containing tetra-ethyl lead as a knock preventive. In the cracking of heavy petroleum oil to yield more gasoline there is formed in the product a considerable amount of what are called unsaturated compounds. These appear to operate to reduce the tendency of the fuel to cause knock in use. The general use of anti-knock fuels will be of unquestionable value in helping to solve the problem of a future fuel supply. If the efficiency of our automobiles can be increased by an average of even 1 per cent, the result will be equivalent to the production of nearly 100 millions of gallons of gasoline per year."

NOVEMBER 1927

"The Norwegians have perfected whaling ships capable of operating in the Antarctic, thousands of miles from their home port. These ships are fitted with a false bow that can be tilted downward into the water to serve as a runway on which one of these huge mammals can be drawn to be cut up. Machinery aboard extracts the oil from the blubber and converts the carcass into meal. These ships are independent of a land base, and having filled their storage tanks with the whale oil, which is in special demand by soap-makers, they steam to whatever world port holds forth the best promise of a profitable market for their cargo. In recent years the number of whaling companies has increased rapidly, and no ocean area is exempt from whaling operations. In excess of 10,000 whales are killed annually, the maximum yield of oil having been reached in 1923, amounting to 44,000,000 gallons. Millions of gallons of whale oil now find a ready market in this country. This freedom of operations without restraint on the high seas has aroused the fear of intelligent observers that whales may soon become commercially extinct. The only possible control of such operations must be found in international agreement."

MAY 1928

"A large percentage of the injuries resulting from automobile accidents are due not to the smash-up of the car itself but to the flying glass from the windows and windshield. Consequently the adoption of triplex glass this year by two well-known automobile manufacturers has aroused great interest. The type of triplex used in the windows and windshields of automobiles is identical in appearance with ordinary plate glass and consists of three layers. The two outside layers are the finest obtainable plate or sheet glass, and the middle layer is a binding composition known as pyroxylin plastic. This is a perfectly transparent sheet of cellulose material. Under ordinary impact triplex will not shatter or create fragments because the flexible center layer holds tightly to the outside layers."

JUNE 1928

"Winging their way westward through storm and fog the three aviators Koehl, von Huenefeld and Fitzmaurice, aboard the Junkers airplane *Bremen*, have made the first successful heavier-than-air east-to-west trans-Atlantic flight in history. To traverse the ocean in the eastward direction is a remarkable feat, but the crossing from Europe westward is fraught with far greater risk, as this German-Irish crew of the *Bremen* knew full well when they attempted it. In the Temperate Zone the weather moves from west to east, rarely from east to west. Steamers make a slower passage when heading westward, and airplanes must negotiate about 600 added air miles in the westward crossing of the Atlantic."

JULY 1928

"Captain George Hubert Wilkins literally leaped to world fame when he flew across the top of the world from Alaska to Spitsbergen in 21 hours. His recent flight, together with evidence provided by some previous deep soundings he had made in the Arctic, reinforces the Nansen theory that the Arctic is a great deep ocean basin, similar in this characteristic to those of the other great oceans of the world."

MARCH 1930

"To Prince Louis-Victor de Broglie, brilliant French physicist, is due the credit for first suggesting the new concept of the wave atom. Since his first publication of this revolutionary concept it has been widely and frequently discussed throughout the world of science. What de Broglie did was to synthesize some of the main concepts of physics that previously seemed contradictory. His hypothesis, as it has been put by the American physicist Paul R. Heyl, 'was that every mass particle was surrounded by a group of waves travelling with the particle as a sort of bodyguard.' Thus electrons, the fundamental basis of matter, have some of the qualities of light; matter becomes less material and light more material, suggesting a kind of merger of previous concepts."

JULY 1930

"The new Lawrence Lowell telescope at the Lowell Observatory in Flagstaff, Ariz., has within a year of its installation brought to light the long-sought Planet X. Photographs of moderate exposure covering the region of the heavens around the predicted position were obtained, and on one of them, taken on January 21, Mr. Clyde W. Tombaugh of the Observatory staff found 'a very promising object.' It was identified from its motion past the numerous fixed stars on plates of the same star field being viewed in the blink comparator. This showed that one faint object among many thousands had shifted its place by a certain expected order of distance in the interval between the taking of the two plates. Since that date the object has been carefully followed both photographically and visually. Its motion in the heavens has been just what might be expected of a trans-Neptunian planet at about the distance predicted by Percival Lowell, and its longitude agrees closely with his predictions. There is no doubt that it is actually a new planet much farther away than any previously known."

SEPTEMBER 1930

"The big news of the recent meeting of the American Chemical Society in Atlanta was the announcement of a new refrigerant, non-inflammable and nontoxic, by Thomas Midgley, Jr. He is the chemist who discovered the anti-knock properties of tetra-ethyl lead. Now he has perfected a refrigerating medium that can be used in household electric refrigerators without the remotest risk of danger from leaky coils. His discovery of the suitability of dichlorodifluoromethane gives promise of the early use of refrigeration for air cooling in homes and theaters and other public gathering places where refrigerating engineers have hesitated to risk accidents with refrigerants that are poisonous or explosive."

DECEMBER 1930

"One of the most important problems engaging the attention of airplane designers today is that of metal covering instead of fabric covering for both the wings and the fuselage of an airplane. Fabric covering has the disadvantage of poor maintenance and durability. Metal covering is unfortunately very heavy, even if the thinnest sheet is employed. Since the airplane is under strict weight limitations, it is only desirable to use metal covering provided the covering itself contributes to the strength. Unfortunately thin metal sheet will not develop anything like its theoretical strength because of local failure termed 'crinkling' when the sheet is in compression. The Navy Department and the Army Air Corps are both giving much attention to the problem of the metal-covered fuselage, and in time efforts of various designers should lead to more accurate knowledge. The large airplane for the future should have a structure similar in some respects to the structure of the ocean liner, in which the metal plating is so thick that it can be confidently taken into account in the calculations of strength."

CHAPTER FIVE

MARCH 1931

"Hermann Oberth of Mediasch, Germany, proposes that we take advantage of the dual forces of gravitation and centrifugal force to send a rocket aloft that will circle the earth continuously. Rising to a height of about 500 miles, equipped with all manner of observation devices, the rocket will acquire a speed of some five miles per second parallel to the earth's surface and then, with its power shut off, will continue to circle the earth indefinitely. From this 'station,' which will circle the earth every two hours, a scientist-observer will watch cloud movements, observe the stars and note changes in the sun, with their resulting effect on terrestrial weather, radio communication and other electric phenomena. As many stations may be established as desired, each, of course, covering a circumferential segment of the terrestrial sphere. Such studies of outer space as are made possible by the station must of course precede attempts to conquer outer space and make a journey to the moon or to one of the other planets."

APRIL 1931

"A new type of stroboscope, a device for 'stopping' motion to study the behavior of machines operating at high speed, has been developed at the Massachusetts Institute of Technology by Harold E. Edgerton. The unique feature of the new instrument is the electrical circuit that causes a condenser to discharge periodically through a thyratron mercury-arc tube. An intense blue light of extremely short duration is produced by a large current through the tube and makes it possible to adapt the stroboscope for photographic as well as visual observation.

So powerful is the new instrument that still and motion pictures of a 160-horsepower electric motor have been made while the machine was running at full speed. Although the rotor was turning over at a rate corresponding to a ground speed of 95 miles per hour, the moving parts were shown as clearly as if the machine were standing still."

DECEMBER 1931

"Eighteen persons stepped aboard a large transport plane of Eastern Air Transport. The chief pilot of the company, Harold A. Elliott, took off from Newark Airport, set his compass course for Washington, threw out a clutch and abandoned his post at the controls. The plane flew on steadily under perfect control for 10 minutes and then Elliott threw in the clutch, turned the plane on a course back toward Newark, again threw out the clutch and let the plane fly with no hand at the controls. In 11 minutes so unerring was the aim and so perfect the control that the plane passed over the center of Newark Airport. This flight was the first public demonstration of the Sperry gyro-pilot, which does everything but take off and land a plane. New possibilities can therefore be seen for a wider and more confident use of airplanes."

MAY 1932

"The most significant scientific discovery of recent years was announced recently in Cambridge, England, where Dr. James Chadwick established the existence of the subdivision of matter known as the neutron. To understand this conception it is necessary to recall that the atom, once regarded as the smallest possible subdivision of matter, is now regarded as being made up of electrons and protons held together by the attraction between negative and positive charges. The neutron is a combination of a single electron and a single proton, the respective electric charges having been neutralized by their union."

FEBRUARY 1933

"The rocket may someday become a very dangerous military weapon. In fact, many authorities think that its utilization in warfare is likely to come much sooner than its application in aerial navigation, and they visualize a powerful rocket guided by wireless and capable of destruction at long range. Perhaps one reason the Germans are so interested in rocket development is that they are debarred from the construction of military aircraft. At any rate, more rocket experimentation is carried on in Germany than in all other countries put together. The latest German rocket design is that of a young engineer named Tilling, of Osnabrück. Tilling recently gave a striking demonstration at the Tempelhof aerodrome in Berlin. The Tilling rocket is made entirely of aluminum and consists of a central shell to which are attached four long tail fins. The over-all length is about nine feet. The rocket motor is inserted into the shell between the fin roots; this motor is filled with a special powder that is relatively slow-burning. As Tilling wished his craft to alight within the aerodrome, he only employed a 12-pound charge in his experiment. The powder was ignited from a distance by means of an electric spark, and the reaction of the exhaust gases drove the rocket upward to a height of about 2,600 feet. Two wings were hidden in the hollow fins until the rocket had reached its maximum altitude. Then a simple automatically acting hydraulic device gradually moved the wings out from their sheaths. The rocket, now converted into a glider, made its descent in a fairly tight spiral. The landing was only 400 yards from the starting point."

JUNE 1933

"When Lt. Comdr. Frank M. Hawks wrote the specifications for the Texaco *Sky Chief*, he thought he was giving the manufacturer a knotty problem. He asked for an airplane that would carry fuel enough for at least 2,000 miles at well over 200 miles per hour, and he was insistent that the airplane have a landing speed not to exceed 70 m.p.h. Heretofore all fast airplanes had a landing speed in a ratio from 2 or 3 to 1, and Commander Hawks was asking the manufacturer to improve that ratio to something like 4 to 1. That the problem had been solved by the Northrop Corporation was immediately obvious after the new *Sky Chief* had been taken up for a few flights. Tentative performance figures showed a top speed of 250 m.p.h. and a cruising speed of more than 200 m.p.h. Performance figures showed a landing speed of from 40 to 50 m.p.h. or a ratio of 5 to 1. The improvement of the ratio was obtained by the use of large wing flaps or air 'brakes.' The lower side of the trailing edge of the wing is split from the upper side for about 20 per cent of the wing chord. Attached to the wing proper by hinges, this flap is raised or lowered by a geared shaft actuated from the cockpit."

JULY 1933

"Invisible 'black light' was described by Samuel G. Hibben of the Westinghouse Electric and Manufacturing Company at a recent meeting of the Illuminating Engineering Society in New York City. 'The "black light,"' said Mr. Hibben, 'is ultra-violet radiation, and it is 99 per cent free of visible light. It is produced by two new black-bulb lamps, one consuming two amperes, the other five. They are made of a special cobalt glass. The bulbs of these lamps absorb 99 per cent of the visible light and transmit 80 to 85 per cent of the ultra-violet wanted. The radiations are relatively long in wavelength, in the range of 3,200 to 4,000 angstrom units. This long-wave ultra-violet lends itself to many photographic and fluorescent effects.'"

SEPTEMBER 1933

"A secret method of depositing photo-sensitive elements on a four-inch square of mica is the foundation of a new television transmission system recently announced by Dr. Vladimir K. Zworykin of the RCA-Victor Research Laboratories. Without mechanically moving parts, Dr. Zworykin has succeeded in transmitting images built up 250 lines to the inch, or 62,500 elements to the square inch. The best that can be done with the conventional scanning disk is about 50 lines or 2,500 picture elements. The modified cathode-ray tube used in the Zworykin transmitting system is called the Iconoscope. The mica mosaic is in such a position that an image can be focused upon it through the glass of the tube. At the base of the tube is an electron-producing gun. The invisible stream of projectiles from the gun is controlled by four coils mounted on a frame and placed outside the tube. When the proper alternating currents from vacuum-tube oscillators are applied to the coils, the beam of electrons can be controlled so that it will sweep back and forth over the surface of the mosaic, covering every portion of it 24 times during each second of operation. The mosaic plate has on one surface 3,000,000 light-sensitive units and on the other surface a plate of silver. Each of the units, together with this plate, forms a tiny condenser that stores up current in exact proportion to the amount of light reaching the unit. This condenser is discharged as the electron beam sweeps over it, and therefore adds its bit of current to the 'picture signal' being built up for transmission. Dr. Zworykin states that this new system will operate at a speed comparable to that of motion-picture camera film. Thus it is possible to televise objects in ordinary light and to dispense with the intense lighting systems that up to now have been necessary. At the receiving end the Zworykin cathode-ray type of producer is used."

NOVEMBER 1933

"In July, Wiley Post startled the world by flying a 15,400-mile ring around it in seven days, 18 hours, 49 minutes. And he did it solo to boot. There was a radio device of paramount interest on the *Winnie Mae*. It was mounted on the globe-circling ship by engineers of the U.S. Army Air Corps at Wright Field. Broadly speaking, it is a direction finder. To make use of it the *Winnie Mae* had an aerial from the rudder post to the fuselage, although she carried no radio in the ordinary sense. Stations along the world route had been asked to broadcast to her in flight on specified frequencies, and this they did with great success. It is obvious that if a pilot can receive signals from two stations of known location while he is in flight, he can orient himself and fix his position. But the direction finder of the *Winnie Mae* did more than this. The signals from a single station actuated a pointer that told Post with great exactitude whether he was on course. It may be said without too much stretching of the truth that he rode a radio wave around the world."

MAY 1934

"With regard to the prediction of earthquakes, seismology has not yet reached the stage where we can foretell quakes ahead of time, but investigations in this direction being carried out in Japan give hope that the time is not far distant when such prediction will be possible. It has been noticed that in earthquake regions the earth shows evidence of tilt or gradual rising for some years before the quake occurs, much as the inner tube of a tire or the bladder of a football rises gradually through a break in the cover before finally bursting. The tilt of the ground is being carefully observed and measured, and it is hoped that it will finally give the clue to the forecasting of earthquakes."

JUNE 1934

"Walter S. Adams and Theodore J. Dunham, Jr. have found that Venus shows no trace of the familiar bands of oxygen or water vapor. But there are three beautiful bands in the deep red and infra-red that have been definitely traced to the carbon dioxide. Such an atmosphere would exert a powerful 'greenhouse' effect, letting the short waves of sunlight in and retarding the escape of the long waves from the warm surface. Rupert Wildt, one of the best men working on the subject, concludes that the temperature at the planet's surface may be as high as the boiling point of water. Life could hardly maintain itself under these conditions, and in its absence the carbon dioxide would remain in the atmosphere. Mars shows no trace of oxygen, water or carbon dioxide. It would seem, then, that Mars may represent a later stage in the history of a planet than the earth, whereas Venus somewhat resembles the earth before life developed upon it. We find ourselves 'wandering between two worlds, one dead, the other powerless to be born.'"

JULY 1934

"One of the major reasons no serious experimenter has yet attempted to build a real moon rocket is the magnitude of the job. Physicists tell us that if we can reach a speed of about 25,000 miles per hour—more exactly 6.664 miles per second—we can then shut off our power. Provided our rocket is outside the earth's atmosphere, it will thereafter coast to its objective on its own momentum. Now, building up a velocity of 6.664 miles per second, particularly in a projectile large enough to carry a crew, will require enormous quantities of energy. If the physicists produce their much-discussed atomic energy, it would help us a great deal. If, as is more likely, we shall have to depend on present materials, the fuels will probably be liquid oxygen and liquid hydrogen or acetylene. These fuels, mingled and fired in a 'blast' chamber, are theoretically capable of giving us the speed we need, provided we can devise a rocket large enough, yet light enough, to carry the cargo."

SEPTEMBER 1934

"The 'synthetic rubber' tire is now an accomplished fact! These relatively insignificant words tell a story of tremendous economic significance. They indicate the successful solution of a long fight to ensure for the United States a source of rubber goods and in particular tires which would make us independent of foreign producers of rubber in time of war. Tires made entirely of Du Prene, the so-called synthetic rubber developed by the Du Pont company, have been built by the Dayton Rubber Manufacturing Company, and severe tests have proved these as tough and durable as tires made of natural rubber."

JANUARY, 1935

"A small electric motor, powered solely from sunlight and able to run continuously as long as the rays of the sun fall on the light-sensitive surface, has recently been constructed by a Detroit manufacturer, J. Thos. Rhamstine. A battery of 20 small light-sensitive and power generating disks, connected together and directly to a small direct current permanent-field motor, turns the motor at a high rate of speed without the use of any source other than the effect of the sun's light rays on the disks. This motor at present has no practical value, for the amount of power produced is very small. But to the scientifically minded it has significance, since the possibilities of obtaining power directly from the sun have scarcely been touched."

JUNE 1935

"The discovery of the isotope of hydrogen, the so-called heavy hydrogen or deuterium, and its successful separation from the light variety, has excited great interest in scientific circles. One problem that challenges physicists is the structure of the atomic nucleus. We understand fairly well the structure of the electronic atmosphere of atoms, that is, the outside structure, but the structure of this minute central sun of the atom is quite unknown. In unraveling the structure of the nucleus, the deuterium nucleus, or deuteron, will certainly play an important part."

MAY 1936

"Electron optics is a comparatively recent branch of the science of electronics. This field of study shows that it is possible to shape electrodes in such a way that the electric field between them will act as an 'electron lens,' which is found to have properties almost identical with those of an ordinary glass lens. A typical image tube with a nine-inch viewing screen has a large-aperture lens mounted to image the scene onto the infra-red sensitive cathode. This image is in turn reproduced on the fluorescent screen of the image tube, thus enabling the observer to 'see' by infra-red radiation, even in darkness. Another use is in connection with infra-red microscopy."

SEPTEMBER 1937

"What is reputed to be the first drive-in bank in the world has recently been opened at Vernon, a suburb of Los Angeles. The motorist merely drives into the bank building and transacts his business without even moving from his seat."

OCTOBER 1937

"A new building code has been adopted by New York City. Among other provisions in keeping with recent progress in the building industry, it permits the use of welding in place of riveting in the construction of steel buildings. This action removes the last serious obstacle to the general use of welding in building construction."

NOVEMBER 1937

"When soybean flour that has been completely freed of oil by solvent extraction is dissolved in water, the resulting solution can be whipped into a stiff white foam greatly resembling egg white. Since protein in the form of soy meal costs only about 1/10 as much as protein in the form of eggs or milk, which it resembles dietetically, research has been directed toward its use to replace these more expensive foods."

MARCH 1938

"Carrots used to be short chubby roots, far less attractive than the long slim beauties seen in many markets today. By careful breeding, a deeper orange color has been developed and the core has been made more tender or practically eliminated. The modern carrot, if well grown, has as little in common with the carrot of the past as a modern streamlined car has with a pre-war gas buggy."

SEPTEMBER 1938

"The details of sun-spots and the like look sharp in the telescope. Yet there is nothing solid there, nothing liquid, nothing even as substantial as a thin summer cloud, with its tiny scattered drops of water—only gas that is hotter in some places than in others. But since this gas is full of electronhaze, we cannot see deep into the sun—probably only a hundred miles or so into the haze. But the haze does not, and cannot, prevent a greater amount of radiation from escaping upward from deeper layers where they are hot than escapes where they are cooler—and so we have the spots."

MAY 1939

"When its temperature is lowered to 2.19 degrees Kelvin, liquid helium seems to stop boiling. This observation is merely one of many that show the temperature marks a transition point between two states of helium. Helium is the warmer, helium II the colder. Helium II possesses an unusual heat-conducting capacity, far in excess of helium I; some time ago Peter Kapitza, the Russian scientist who spent many years in Lord Rutherford's laboratory at Cambridge, England, thought this might be due to an abnormally high fluidity—or low viscosity, as the physicists call it. This would mean that the liquid would circulate with extreme ease, carrying heat by means of convection currents. The viscosity of helium II now turns out to be about the same as that of gaseous hydrogen, low enough to set the theoretical physicists to try to explain these superfluid properties."

OCTOBER 1939

"If atom smashing could be made more efficient, power production by means of nuclear fission would not be beyond the realm of possibility. But under present conditions, the process is as inefficient as removing the sand from a beach a grain at a time. The secondary neutrons emitted at the moment of fission and in later reactions, however, bring up an interesting and rather disturbing aspect of the case. These secondary neutrons constitute a fresh supply of 'bullets' to produce new fissions. Thus we are faced with a vicious circle, with one explosion setting off another, and energy being continuously and cumulatively released. It is probable that a sufficiently large mass of uranium would be explosive if its atoms once got well started dividing. As a matter of fact, the scientists are pretty nervous over the dangerous forces they are unleashing and are hurriedly devising means to control them. It may or may not be significant that, since early spring, no accounts of research on nuclear fission have been heard from Germany—not even from discoverer Otto Hahn. It is not unlikely that the German government, spotting a potentially powerful weapon of war, has imposed military secrecy on all recent German investigations."

FEBRUARY 1940

A plant for the commercial manufacture of nylon yarn, erected by the du Pont Company at Seaford, Delaware, went into production on December 15 last. Nylon hosiery will be made by a number of nationally known hosiery manufacturers, and it is anticipated that nylon hosiery for both men and women will be put on the general market by late spring or early summer of 1940. The manufacturers say that the word 'nylon' has no significant derivation; that it was selected as a generic name because it is non-technical and is easy to pronounce; its individual letters do not stand for anything."

MAY 1940

"Illness is industry's biggest bill! In the heavy industries, employing some 15 million persons, the average male worker loses about eight days a year because of sickness. Weary workmen are easy marks for illness and accidental injuries. One Pennsylvania manufacturer instituted two 15-minute rest periods, one in the morning and one in the afternoon. Although this shortened the work day by half an hour, production jumped by 20 percent."

JUNE 1940

"In these parlous times we must perforce be interested in such gentle arts as dive bombing. The German attack on Poland showed the value of the method in attacking tanks, and generally in support of infantry. The speed of the dive increases the vertical velocity of the bomb so that its penetration is equal to that of a projectile released in level flight at a great altitude. Therefore, in spite of the fact that the dive bomber is peculiarly vulnerable to shell fire at low altitudes, dive bombing must remain a powerful aerial tactic."

NOVEMBER 1940

"Oxygen in the ocean is probably used up more rapidly by bacteria and other micro-organisms than it is by all the fish and other visible animals ranging from tiny shrimp to giant octopuses, suggests Dr. Claude E. ZoBell of the Scripps Institution of Oceanography at La Jolla, California. A quart of ocean water may contain anywhere from 100,000 to 10,000,000 bacteria, consuming oxygen at the rate of .001

of a cubic centimeter to more than one cubic centimeter per quart per year."

JANUARY 1941

"Hunters and skeet shooters 'lead,' or aim ahead of their targets, but their guns are fired from a nearly stationary rest. Aerial huntsmen have to bag 400-mile-per-hour pursuits from ships traveling at comparable speeds. A gunner firing from a plane in a side-slip, for example, must allow for the fact that his weapons are traveling through *three directions* simultaneously; forward, side-wise, and downward. To facilitate aim, gunners employ sights which swing off center to compensate for the effect of the slipstream. There is, however, another phenomenon called the 'Magnus Effect' for which no sights will compensate. It is induced by the clock-wise spin that is imparted the bullets as they leave the gun barrel. When the gunner swings his armament for a left broadside and fires, the spin causes the bullets to drop in their flight just as top-spin drops a tennis ball. Contrariwise, from a right broadside, the bullets would rise in flight like a golf ball given backspin."

MARCH 1941

"The two wolf-children of India were first seen living as wolves among wolves on October 9, 1920, by an Anglican missionary, Rev. J. A. L. Singh. Rev. Singh and his wife expected that a few years of association with the normal children in their orphanage would change the wolf-children from effective little animals back into human beings. They were to be disappointed. The children hated, feared, and shunned human beings, as would have a wolf-cub. Kamala, the elder surviving girl, gradually developed into a pathetic little, sub-normal, but clearly not idiotic, human being. She learned to speak about 50 words and occasionally to put them in short sentences. From the entire account it becomes clear that, while the normal baby is born with the potentialities to become a human being, man actually attains this only through association with his kind in the very earliest years."

JUNE 1941

"A skin test which tells in less than an hour whether or not a woman is going to become a mother has been announced by Dr. Frederick H. Falls, Dr. V. C. Freda, and Dr. H. H. Cohen, of the University of Illinois College of Medicine. The test is similar to those made for allergy to hayfever and is said to be 98-percent reliable. Colostrum, a watery liquid secreted in the breasts during pregnancy until milk formation starts, is injected by hypodermic needle into skin of the forearm. If the woman being tested is pregnant, there is no reaction. If she is not pregnant, a reddish area of one or two inches in diameter appears within an hour around the injection point."

JULY 1941

"Large-screen pictures constitute the most dramatic development of television in recent days. At an RCA and NBC demonstration in a New York theater, a complete program was presented utilizing equipment that had been specially installed for the purpose. The program consisted of a play-let put on by 'live' actors, a news broadcast, a newsreel picked up from film, and a ring-side pickup of a championship boxing match. Spectators stated that the quality of reproduction was good, and that the views of the fight were better than could be had from a ring-side seat."

SEPTEMBER 1941

"A record of nearly 10,000 American children brought into the world with the aid of the proxy-father procedure—technically termed artificial insemination—is reported in the *Journal of the American Medical Association*. The central and Atlantic seaboard sections of the United States have the greatest number of children sired by artificial insemination. More than 97 percent of all the pregnancies resulted in living normal babies. The number of miscarriages and abortions was only one fifth the usual rate."

NOVEMBER 1941

"The pilots who fly the big bombers over the Atlantic to Britain sometimes stay at 15,000 feet, or higher, the whole way over. It's 50 degrees below zero up there but the heated planes are comfortable. Insidious, though, is the effect of altitude: at first you don't recognize the dreamy, don't-care feeling as the higher centers of the brain gradually cease functioning and you may wait too long before attaching the oxygen tube. One pilot, flying in winter at 20,000 feet to avoid icing, had to detach his tube and go back to help a passenger. When he returned to his seat he had to readjust his tube, a simple operation. But he couldn't do it. The tube in his hand approached the socket—wavered away. That went on for five minutes while they slipped down toward the dangerous icing level. Then the navigator realized what was wrong and came to the rescue."

APRIL 1942

"The case against flies as the culprits that spread infantile paralysis is strengthened by a discovery reported by Dr. Albert B. Sabin and Dr. Robert Ward, of the Children's Hospital Research Foundation and the University of Cincinnati College of Medicine, in the current issue of *Science*. Previous discovery of the infantile paralysis virus in flies was made in rural areas. Flies caught in Cleveland and Atlanta in the neighborhoods of infantile paralysis patients were infected with the virus of the disease. Discovery of the virus in city flies is considered more significant, especially since the infected flies were found in modern neighborhoods with good plumbing and in which several children had mild illnesses that might have been abortive infantile paralysis."

JUNE 1942

"Engineers at the Westinghouse high-voltage laboratory recently caught man-made lightning in a bucket of sand to produce replicas of the glass-like fulgurites formed by natural lightning strokes. 'Since a temperature of about 3,000 degrees is required to melt sand into fulgurites, these experiments give us definite knowledge of the tremendous heat which lightning can produce,' Dr. P. L. Bellaschi, directing the experiments, reported. 'Fulgurites might be called petrified lightning, since they have the same crooked shape as the bolts that formed them. Natural specimens occasionally are found buried in the ground, particularly in dry desert sands. They are glass-like tubes of solidified sand, formed when lightning surges through dry earth in search of moist ground in which to neutralize its charge."

AUGUST 1943

"Astonishing 30-day cures of long established hives cases resulted from oral administration of a drug which neutralizes histamine. The same drug relieved the histamine-sensitive patients of skin eruptions and acid stomach. Even rheumatoid arthritis and swelling of the legs and arms have been benefited. Dr. Louis E. Prickman, of the University of Minnesota, believes that antihistamine therapy offers great possibilities in the correction of food allergies."

SEPTEMBER 1943

"A new antibacterial substance, penicillin, has joined the ranks of the 'miracle drugs.' Clinical tests of the material give good reason for belief that it is superior to any of the sulfonamides in the treatment of *Staphylococcus aureus* infections. Preliminary tests on wounds and infections of soldiers returned from the battlefronts have been so encouraging that the tests are going forward on a broad scale. In this work many difficulties are encountered. They arise chiefly from the facts that the mold, *Penicillium notatum*, from which penicillin is obtained, produces only tiny amounts of antibacterial substances after a long period of growth in a culture medium that must be very carefully protected and controlled. According to a recent report, a yield of as much as one gram of purified penicillin from 20 liters of culture fluid would be an excellent result."

NOVEMBER 1943

"Glass with non-reflecting surfaces, developed for military uses by American Optical and RCA, can be applied, with desirable results, to post-war manufacture of many useful items. Among the new products are windshields sans dangerous reflections, less conspicuous spectacle lenses, more easily read instruments, faster camera lenses, shop windows free from reflections, more efficient microscopes and other light-transmitting instruments."

JULY 1944

"In many war plants, workers may be seen tapping objects, one after another, in front of a microphone. Little or no sound can be heard by human ears, yet every now and then a light flashes and the operator tosses a piece aside as defective. This is just one of several new techniques which utilizes supersonic frequencies for inspection purposes in industry. Cracks, differences in hardness, changes in dimensions, and variations in the composition of many materials can be quickly detected by this method."

AUGUST 1944

"Engineers for years have sought a practical method of gasoline injection for supplying fuel to the cylinders of gasoline engines. Such a method has now been perfected and is in production, according to Donald P. Hess, President of American Bosch Corporation. 'The gasoline, by this system, is delivered uniformly to every cylinder of the engine. The result is that all cylinders pull together in harmony, producing a smoother flow of power and quieter engine operation than has ever been possible with any other method,' Mr. Hess states."

DECEMBER 1944

"A new synthetic foam rubber, as soft and fluffy as an angel food cake, has been announced by The Firestone Tire and Rubber Company. Whipped into a creamy froth, much as a housewife beats egg whites for her cake, the synthetic latex traps innumerable interconnected tiny air bubbles, which give the foam rubber its softness and permit free circulation of cooling air."

MAY 1945

"A recent development in plastics and electronics is a wafer-thin Vinylite plastics record, only seven inches in diameter. Each side of the disk will record approximately 15 minutes of dictation. These records can be bent, rolled, dropped, and written on with a pencil without harming the sound track. The thin plastic can be stored indefinitely, without warpage, breakage, or distortion, in an ordinary filing cabinet—100 disks to the inch—and played back at least 100 times."

NOVEMBER 1945

"Miniature oxygen tents for babies born prematurely are now being fabricated from Ethocel sheeting. Still in the experimental stage, the clear plastic tents permit a full view of the tiny patient."

JUNE 1946

"Capable of solving scientific problems so complex that all previous methods of solution were considered impractical, an electronic robot, known as Eniac—Electronic Numerical Integrator and Computer—has been announced by the War Department. It is able to compute 1,000 times faster than the most advanced general-purpose calculating machine, and solves in hours problems which would take years on a mechanical machine. Containing nearly 18,000 vacuum tubes, the 30-ton Eniac occupies a room 30 by 50 feet."

JANUARY 1947

"It is really astonishing to find what effects odors can have on purchasers. A case in point: scented hosiery is bought in preference to unscented hosiery, but, oddly enough, a survey has shown that purchasers are not consciously influenced by the odor; they imagine that the scented goods have a better texture or a more appealing color."

APRIL 1947

"The new camera of Edwin H. Land, founder and president of the Polaroid Corporation, is appraised by experts as one of the greatest advances in the history of photography. The Land camera is similar in many respects to the ordinary camera. However, after you snap your picture, exposing a section of film in the ordinary way, you turn a knob which pulls a length of film and printing paper through a slot to the outside of the camera. Glued across the paper, at intervals representing the length of one print, are a series of narrow, metal-foil envelopes, or 'pods,' each containing a quantity of a thick, sticky paste. As you turn the knob, the little 'clothes wringer' squeezes open one of the pods, and the paste is spread evenly between the negative and the paper. The sandwich now in your hand is a miniature darkroom. You wait for about one minute, then you peel apart the layers, and there is your finished picture, neatly framed in a white border."

SEPTEMBER 1947

"Reported to have a lower thermal conductivity than still air, heretofore theoretically considered the most efficient thermal insulator, a new material is 6 percent silica and 94 percent air. Chemically known as an aerogel, this new insulator is so efficient that it will make possible an increase in refrigerator and freezer capacity of up to 60 percent."

APRIL 1948

"A new X-ray gage measures the thickness of red-hot steel without physically contacting it in any way. The device shoots one X-ray beam through the hot steel strip as it moves off the finishing stands in a rolling mill. Simultaneously, a second X-ray beam from the same source penetrates a standard reference sample of a desired thickness. The instrument then compares the intensity of the two beams; a difference indicates that the strip is either more or less than the desired thickness."

JULY 1948

"Nineteen years after Edwin Hubble's discovery that the galaxies seem to be running away from one another at fabulously high speeds, the picture presented by the expanding universe theory—which assumes that in its original state all matter was squeezed together in one solid mass of extremely high density and temperature—gives us the right conditions for building up all the known elements in the periodic system. According to calculations, the formation of elements must have started five minutes after the maximum compression of the universe. It was fully accomplished, in all essentials, about 10 minutes later."

SEPTEMBER 1948

"Within the past few months a group of physicists at the Bell Telephone Laboratories has made a profound and simple finding. In essence, it is a method of controlling electrons in a solid crystal instead of in a vacuum. This discovery has yielded a device called the transistor (so named because it transfers an electrical signal across a resistor). Not only is the transistor tiny, but it needs so little power, and uses it so efficiently (as a radio amplifier its efficiency is 25 per cent, against a vacuum tube's 10 per cent) that the size of batteries needed to operate portable devices can be reduced. In combination with printed circuits it may open up entirely new applications for electronics."

MAY 1949

"On February 24, 1949, man made his first really substantial step into outer space. An ex-German V-2 rocket took off from the White Sands Proving Ground in New Mexico. In its nose it carried an American-made rocket, the Wac Corporal, filled with telemetered instruments. The Wac Corporal began to burn its own fuel at an altitude of 20 miles and coasted upward to an altitude of 250 miles. While definitions of the limit of the atmosphere differ, it is fair to say that at the peak of its ascent the Wac Corporal was in interplanetary space. The largest promise in that shot was in the use of the step (now called stage) principle. If the principle can be extended to three steps, we may get a 'satellite rocket' that will circle the earth."

JUNE 1949

"Two University of Chicago physiologists, Ralph W. Gerard and Robert T. Tschirgi, have succeeded in keeping a large section of a rat's spinal cord alive and functioning outside the animal's body. Placed in a trough after dissection, it is supplied with blood or an artificial nutrient through the spinal-cord arteries. Gerard and Tschirgi have already found five distinct substances capable of furnishing energy for nerve. (Glucose had previously been considered the only energy source.) They have also been able to demonstrate that spinal-cord function—in apparent contrast to accepted theories of brain function—can be restored after as much as 30 minutes of oxygen or glucose deprivation."

CHAPTER SIX

NOVEMBER 1846

"If there is any one crime which should excite universal indignation, it is the sneaking villainy of cutting the wires of the magnetic telegraph. This scoundrelism, if not checked by the vigilance of the whole community, appears likely to deprive the public of the important benefits to be derived from this greatest invention of the age. It is supposed by some that this mischief proceeds from sheer envy against the rapidly advancing honor and prosperity of our country, under a system of free institutions and unbridled enterprise."

DECEMBER 1847

"A four-wheeled carriage with brown ornaments and iron wheels has been recently discovered in a three-story house dug out at Pompeii. It is our opinion that when the Roman Empire was overthrown by the Goths, the Romans were nearly as far advanced in civilization as we are at the present moment."

JULY 1846

"Simultaneous and instantaneous ignition of gas lamps in towns by means of electricity, states a correspondent, will ere long be substituted for the present slow and irregular method. He further states, 'I confess that I am astonished that electricity has never been enlisted into the service of the steam engine, when every clear intellect must perceive that it must ultimately do away with the present employment of fuel and boilers, and their auxiliaries.' "

FEBRUARY 1848

"The Nantucket Enquirer draws a discouraging picture of the prospects of the whaling business in that place. Since the year 1843 the whaling business has been diminished by fifteen sail, by shipwreck, sales, etc. The voyages are said to be one third longer than they were twenty years ago, and the number of arrivals and departures is constantly growing less and less. The consumption of whale oil has been decreasing for a long time as well as the supply. Other carbonic materials are now applied to purposes for which fish oil at one time was alone used."

JUNE 1848

"Miss Maria Mitchell, of Nantucket, discoverer of the Comet which bears her name, was unanimously elected an honorary member of the American Academy of Arts and Sciences, at their last general meeting. We believe that this is the first time such an honor has been conferred on any lady in this country; and a similar honor had been conferred on but two ladies in Europe, Miss Caroline Herschell, the sister and assistant of the late Sir William Herschell, the astronomer, and Mrs. Mary Fairfax Somerville, the commentator on Marquis de La Place's Celestial Mechanics."

OCTOBER 1848

"An average of a million dollars' value is annually wrecked on the Florida Reefs and Keys, for the want of an accurate chart of that coast. Although Florida has been held by the United States for twenty-seven years, no original American chart has ever been made of its dangerous coast. Navigators have to depend upon old Spanish charts, and those made by the British from 1763 to 1784."

FEBRUARY 1849

"Dr. Jones, of this city, proposes to run telegraph wires from St. Louis, Missouri, with a branch to Behring's Straits, where the wires should cross to the Asiatic side, and proceed through Siberia to St. Petersburg, and the principal cities of Europe. In such a project, the governments of Europe, Russia at least, will not be likely to engage—the language of freedom would too often travel along the iron wings to suit the policy of a one man government."

JANUARY 1850

"It is the fortune, or misfortune, of every age, we cannot tell which, to be the witness of great events that never transpire. In the line of navigating the aerial ocean above us, how many triumphant lucky inventors have arisen, some to delude themselves, and some to delude others. During the past two years, in London especially, and from there to the ends of the world, nothing was heard of from time to time but the great 'Electric Light.' A Frenchman discovered one kind, an Englishman another, and a Scotchman another; all were to make short work of gas companies. At one time the price of stocks fell considerably, and there was no little panic in the gas market. It has turned out after all, that the old kind still maintains its position, while the jeers of its younger opponents have been converted into an expiring moan. The steam engine, the steamboat, and many other good inventions, had once supreme judges of wiseacres, who wagged their heads in portentous dignity at the folly and credulity of man."

AUGUST 1850

"The men of forty-five years of age, now living in our city, have seen the first successful steamboat which navigated our waters, and the young man of twenty-one, he who has just arrived at the age of manly responsibility, is a contemporary of the first locomotive. What revolutions these two inventions have produced—steam navigation and railway locomotion—and what a gorgeous panorama passes before our vision as we trace the progress of other inventions. In 1809 there was only one steamboat in the whole world; now, who could count their number? In 1830 there were only thirty miles of locomotive railway in the world; now there are no less than 18,000 miles.

What with the steamboat, the railroad and the telegraph, the ends of the earth are brought together, and civilization is now fast finding its way into the most darkened corners of the earth. And shall it ever be that we shall see the atmosphere as safely navigated as we now see the ocean?"

APRIL 1851

"The locomotive is the most perfect of machines. It approaches nearer to the spiritual and physical combination of the human machine than any other. In it we behold what the steam-engine is when 'unchained to the rock, and unfettered to the soil.' One of the grandest sights in the world is a locomotive with its huge train dashing along in full flight. To stand by night at the side of a rail-road, when a large train is rushing along at the rate of 30 miles per hour, affords a sight both sublime and terrific. No wonder the simple backwoodsman declared that the first locomotive he ever saw was 'pandemonium in harness.'"

JULY 1853

"A terrible riot occurred on June 22nd at the residence of Dr. George A. Wheeler, of New York, caused by the finding of some human bones on the premises. A mob of 3,000 collected, armed with clubs, axes and stones. Dr. Wheeler's store and dwelling were attacked, the inmates driven out, and the premises completely gutted. Not one of the mob who had his arm or leg broken, but would run or get carried to a doctor to get it set, and how could he do this unless he was acquainted with the anatomy of the human body?"

MAY 1854

"Prince Paul of Württemberg is now in this country collecting botanical and ornithological specimens for the publication of a work when he returns to Europe. This is a very creditable occupation for a Prince, and it would be more to the honor of them all if they engaged in some such useful and instructive profession."

JUNE 1854

"Since the fall of the Wheeling Suspension Bridge, articles have appeared in a number of our daily papers condemnatory of iron as a material for such structures. There can be no doubt, in our opinion, but the breaking down of so many iron bridges in our country can be traced to the bad quality of iron used in their construction—it did not possess sufficient elasticity. All iron is iron, just as all wood is timber; but there are just as many varieties of the former as of the latter. Yet how small is the amount of knowledge possessed by the most experienced engineers of the different kinds of iron in comparison with our knowledge of wood. Let civil and mechanical engineers look more to the quality of the iron which they use for various purposes, and the community will not be so often inflicted with painful accidents on sea and land—from the bursting of boilers, the fracturing of the shafts and beams of engines, and the breaking down of iron bridges."

APRIL 1857

"From the number of communications which we have received on the 'divining rod,' we cannot question the honest belief of a number of our readers in its virtues. There are many phenomena in nature which are yet sealed up to us, and the divining rod may be one of these; still, we must say that we are skeptics in the powers or virtues which are attributed to it. We believe that any man of a reflecting and observing mind can guess where water may be obtained by boring, without a divining rod, as well as another person with one. Our opinion may be wrong, but we cannot come to any other conclusion by reasoning on the subject from scientific data. If, however, we are at any period of time after this convinced by ocular demonstration that there is scientific virtue in the divining rod, we will frankly make the change of our views known."

JULY 1857

"Some time ago an offer of $500 was made through the Boston Courier to any one who could exhibit in the presence and to the satisfaction of certain professors of the natural sciences at Harvard University any such marvelous phenomena as were commonly reported by spiritualists as having transpired through the agency of 'mediums.' This challenge was accepted, and several persons professing to have spiritual communications met in the Albion building, Boston, to show their powers. Among the number were the 'Fox girls,' celebrated for their achievements in this line. The committee appointed to judge the case consisted of Professors Peirce, Agassiz, Gould and Horsford. The spiritual experiments were conducted for several days, and the medium allowed ample and fine opportunities of making demonstrations; but like the priests of Baal, in the days of Elijah, they failed to call down their deities."

JUNE 1858

"A few years ago, some members of the Royal Institution of Great Britain suggested that could a telescope be placed at a great elevation, say 10,000 feet above the level of the sea, the observer would be able to scan a greater distance than had ever been seen before. In accordance with this suggestion, Professor C. Piazzi Smyth, Astronomer Royal for Scotland, amply provided with instruments, went to the Peak of Teneriffe, and the history of his observations he has just made known. Professor Smyth was accompanied by his wife, and a better assistant no astronomer ever had. They bivouacked on the top of Mount Guajara, 8,900 feet high, where the air was always calm, the temperature averaging 65 degrees, far above all clouds, and under a sky gloriously resplendent with stars. We have no doubt but that an observatory will be established on Teneriffe by some people or nation. We should like such a place to be cosmopolitan—open to the astronomers of the world; for the stars shine alike on all men, and know no distinctions of flags or nationalities."

OCTOBER 1858

"On almost any pleasant day a portly man with flowing hair, white cravat, and broad-brimmed Kossuth hat, may be seen on Broadway, dashing along behind a splendid pair of fancy horses, fit for the stud of an emperor, and with all the ease and indifference of a millionaire. That man is Elias Howe, Jr., once the poor and humble inventor, who in 1849 obtained a patent for the first practically useful sewing machine. We rejoice in the good fortune of our old friend, and can only say to him that he is entitled to all that he has received."

MAY 1859

"It has been proposed that a by-wash should be built at every dam on our rivers and creeks once frequented by salmon, for the purpose of allowing then to pass up to old spawning grounds. With the artificial cultivation of young salmon, and dams formed with shutes up which the salmon might run to spawn, we have no doubt but the Merrimack, Connecticut and Hudson River would once more abound with the delicious fish."

OCTOBER 1859

"The expedition fitted out two years ago, under Capt. McClintock, at the expense of Lady Franklin, to search for her husband in the Arctic regions, has returned with full and correct tidings of the sad fate of Sir John Franklin and his companions. Captain McClintock found the record and remains of Franklin at Point Victory; and it seems that he died in June, 1847—about 11 years ago. The whole of his companions also perished, some at one place and some at another, in those inhospitable and desolate regions. We hope the last expedition to these dread solitudes of ice and snow has been made."

NOVEMBER 1859

"Hitherto the public parks of New York City have been so small as to excite the derision of foreigners; but this city has at last nobly redeemed itself in this respect by the purchase and arrangement of the 'Central Park,' which, when completed, will be one of the largest and most beautiful in the world. The Central Park of New York embraces an area of 843 acres, and is two and a half miles long by half a mile wide. Three artificial lakes are being laid out, one of which covers 20 acres, and forms the skating pond. A large extent for fields and lawns is assigned for the evolutions of military companies, games of ball, cricket and other athletic sports."

JANUARY 1860

"Dr. Hiram Cox, the Cincinnati inspector, has found that in 700 inspections of stores and lots of liquors of every variety, 90 per cent were impregnated with the most pernicious and poisonous ingredients. Nineteen young men were killed outright by only three months' drinking of these poisoned liquors. Many older men, who were only moderate drinkers, died of delirium tremens. Of 400 insane patients, he found that two thirds had lost their reason from this same liquor. One boy of 17 was made insane from being drunk only once. Seeing two men drinking whiskey that was so strong it actually caused tears to flow from the eyes of one of them, the doctor obtained some for his tests. He found it to contain only 17 per cent of alcohol, when it should have contained 40, and that the difference was supplied by sulphuric acid, red pepper, caustic potash and strychnine. A pint of this liquor contained enough poison to kill the strongest man."

FEBRUARY 1860

"A very distinguished English inventor has recently gone the way of all the earth, and under circumstances which call forth our sympathies. This man was James Boydell, the inventor of the steam traction engine for common roads. He was in the very prime of life, and had made an appointment with some of the government snobs to test his engine in the Woolwich marshes, and was kept waiting in vain for their attendance during a severely inclement day in December, by which he caught a severe cold, from the effects of which he never recovered. His peculiar locomotives are not adapted for quick traveling, but for drawing heavy loads, such as would require 10, 12 and even 20 horses. They are invaluable, and will yet come into more general use. English papers complain that James Boydell was murdered by government routine."

MARCH 1860

"The workers in stone and marble in our cities are subject to lung diseases by inhaling dust. Their average age is 38 years, and they are very frequently sick. We have known several who had to quit the business in New York on account of threatened consumption, who were otherwise of powerful frames and naturally robust. Those who grind the knives and forks with which we eat our food are very short-lived. The dust which is inhaled by the lungs soon coats them with stone. Wet grinding, such as that of saws, axes, knives, is not so unhealthy, but the dry polishing of every tool greatly shortens the lives of those engaged at the business. An exhaust fan to draw off the dry dust should be applied to every emery wheel in factories under a penalty, and all those engaged in these operations should never permit a razor to be drawn across the upper lip."

JUNE 1860

"The advent of the Japanese Embassy, and the interest in this but partially known people, has induced us to give some account of their achievements in the agricultural and mechanical departments. Being compelled to make the most of their not very extensive and rather poor soil, they have arrived at a very high state of perfection in the arts of agriculture. Where the land is inaccessible to the plow, it is cultivated by manual labor. Like the Chinese, they pay great attention to manuring and to irrigation. The short time the Embassy has already been with us shows how eager they are to profit by the experience of foreigners, and to imitate their useful arts. The inhabitants of Japan are already supplied with microscopes, telescopes, clocks, watches, knives, spoons, etc., made by themselves from European models. They manufacture Colt revolvers and Sharps rifles, and it is said that they have made improvements upon them. At Nagasaki works have been erected for the manufacture of steam engines without European assistance."

SEPTEMBER 1860

"Of all inventions of which it is possible to conceive in the future, there is none which so captivates the imagination as that of a flying machine. The power of rising up into the air, and rushing in any direction desired at the rate of a mile or more in a minute, is a power for which mankind would be willing to pay very liberally. What a luxurious mode of locomotion! What little attention this subject has heretofore received from inventors has been almost wholly confined to two directions: flying by muscular power and the guidance of balloons. Both of these we have been accustomed to regard as impracticable. The thing that is really wanted is a machine driven by some natural power, so that the flyer may ride at his ease. For this purpose, we must have a new gas, electric or chemical engine. The simplest of all conceivable flying machines would be a cylinder blowing out gas in the rear, and driving itself along on the principle of the rocket. We might add several other hints to inventors who desire to enter on this enticing field: but we will conclude with only one more. The newly discovered metal aluminum, from its extraordinary combination of lightness and strength, is the proper material for flying machines."

OCTOBER 1860

"Nearly all of our microscopists, in their communications to Silliman's *American Journal of Science* and other kindred works, use the millimeter as their measure, and in Cooke's Elements of Chemical Physics and other standard works, the meter and kilogramme, as well as the degrees of the centigrade thermometer, are employed without translation. We are beginning to think seriously of adopting this course in SCIENTIFIC AMERICAN. The people are running ahead of our legislators in making this great reform in our weights and measures."

APRIL 1862

"The insurance companies in London, like those in New York, have become alarmed at the large quantity of well oil at present stored in the British metropolis. These companies have laid their grievances before the Mayor, and they assert that this oil is of a most inflammable and dangerous character, being liable to spontaneous combustion. It is said that there are about half a million gallons of such oils now stored on the wharves in London. As crude petroleum is more dangerous than the refined quantities, and as the cost for carriage to market is just the same for both, it would be well to refine all petroleum in the vicinity of the oil wells."

SEPTEMBER 1862

"Since the experiments of Mr. Joule in obtaining a mechanical equivalent for a unit of heat, by proving that the temperature of a pound of water will be raised by 1° by the same quantity of power that will raise a weight of one pound 772 feet high, the theory that heat is a condition of matter and not a substance is more generally admitted than formerly. But the popular idea of a material heat is one from which it is very difficult to disembarrass the mind. But as the mind becomes familiar with the idea of heat as a sensation, the various changes of matter daily occurring in nature can all be satisfactorily viewed without recurring to the notion that any invisible substance is entering or leaving the particles."

AUGUST 1866

"The published scale of prices of the Atlantic Telegraph Company shows that for a message of 20 words, including date and address of sender, the sum of £20 will be charged-which is equal to $150 American money at the present rate of gold. Further than that all figures must be written out, and they will be charged as words. Messages in cipher will be double the above rates. Vast amounts of money have been invested and sunk in laying the cable, and its permanency is at least uncertain, but it does not seem to us judicious to attempt to get all the money back this summer. There are not many journals or firms that can afford to have regular messages of any length, and under the circumstances the news transmitted would be scanty and indefinite. Heavy rates defeat the end and aim of such enterprises, which are to be a popular medium for the transaction of business. The cable, however, is not indispensable; steamships cross in nine days—from land to land in much less time—and except in cases of great urgency the capacity of the line will not be taxed to its utmost, unless the tariff of charges be considerably reduced. Doubtless the competition of the Collins Overland Telegraph will have a healthy effect and aid materially in lowering the price."

SEPTEMBER 1866

"A correspondent, A. J. H., writes on the navigation of the air, insisting that all that is required to make it a permanent success is a proper motor, which will combine the necessary power with the requisite lightness, and says that steam is that motor. He claims to have made a rivetless boiler, which will bear a pressure of from 1,000 to 3,000 pounds per square inch. By thus condensing an enormous power he believes aerostation is an accomplished fact or at least is possible. We cannot agree with him that steam, however much super-heated, is adapted to the purpose. The weight of water, fuel and machinery, to say nothing of the boiler, will be found to be too great, when compared with the mass to be moved in a fluid such as air, to have much margin for available power. What is needed, not only for aerostation but for other purposes, is an entirely new motor which shall dispense with the weight of a boiler with its necessary appurtenances. That a new motor will in time be contrived, without these drawbacks, we have no doubt; but until it is done we have but little faith in economic and successful navigation of the air."

JANUARY 1867

"Dr. Jenner in his remarks on Nov. 12th at the opening session of the London Epidemiological Society, of which he is President, advocated the introduction of sanitary science as a regular part of a liberal education. We would go further and urge its adoption as an element of common-school education in its simpler laws and principles, and in its philosophy as an essential of professional education, equally with chemistry, for example. Dr. Jenner's arguments are abundantly forcible for our conclusion. The difficulty in the present state of general education of spreading practical sanitary knowledge and of inducing men to act so as not to destroy themselves and their neighbors is all but insuperable. Constant and indefatigable iteration on the part of the few—line upon line, precept upon precept, example on example, warning on warning—offer the only hope of gradually awakening and instructing the present generation with regard to the common laws of health and disease. The next generation might be and should be better indoctrinated. Mean-while every press and every public instructor of whatever kind should give prominence to the daily lessons of experience and science on this all-important subject."

JANUARY 1867

"In 1866 the expenses of the city of Paris amounted to $46,000,000. In return for this seemingly large expenditure the Parisians had the cleanest and best governed city in the world, together with an astonishing development of great improvements, in the opening of broad spacious streets and in the erection of splendid public buildings. New York City expends about $18,000,000 and gets in return dirty streets, a brutalized swindling political ring and no improvements that are worth mentioning. During the past 10 years enough money has been stolen from our burdened tax-payers to have furnished this city with museums, art galleries, monuments, etc., that would have attracted the attention of the whole world."

OCTOBER 1867

"Whenever a boiler explosion occurs, the attention of the coroner's jury is directed to the discovery of imperfection in material or workmanship. Frequently the engineer and fireman, or the individual who combines both these offices in himself, is removed by the explosion from all opportunity to give his testimony, and the proprietors are unable, if not unwilling, to give light on the subject. Sometimes the engineer or fireman is censured, but seldom is the employer reprimanded. A correspondent says that it is surprising there are not more explosions. He says that in Connecticut the engineer is often required to be his own fireman, to do every 'chore' in and around the whole establishment for $1.50 to $2 per day. He asks, 'Who is going to study and fit himself for an engineer with such remuneration and such duties before him? Why doesn't our State Legislature make a law prohibiting any one from running an engine who is not a competent engineer?"

APRIL 1869

"A good needlewoman with her needle makes from 25 to 30 stitches per minute, whereas a modern sewing machine will make 1,600. Yet we cannot call this last a *labor-saving* machine, so far as regards the operator on it. As compared with sewing by hand, sewing by machine is really a very laborious and fatiguing occupation. Our stage and street horses are changed several times a day, but sewing girls at their machines are expected to work for 10 or 12 consecutive hours with intermittent but continually repeated motions of the muscles of the lower limbs. We have before us a very interesting report on the sanitary condition of the many sewing machine operators who came under his personal notice in the public hospitals of Paris. Hollow cheeks, pale and discolored faces, arched backs, epigastric pains, predisposition to lung disease and other special symptoms too numerous to be specified were found to be the general characteristics of all the patients. These disastrous effects must eventually tend toward the deterioration of our race and deserve, from a humanitarian point of view, the most serious consideration of all friends of mankind."

JULY 1869

"The public is watching with great interest the attempts making to promote industrial and scientific education in this country. Undoubtedly the most important effort of this kind is the Cornell University at Ithaca, N.Y. The tide of opinion has of late been rapidly setting toward a more practical kind of education than has for a long time prevailed. The applications of scientific discovery have revolutionized the arts, and success in any department of industry is getting to depend more and more upon knowledge of fundamental principles. It has therefore become necessary to provide for the special education of youth in order to fit them for anything like a high station in any industrial department. Such a conclusion could not long be entertained with-out attempts to put it into practice, and schools have been established in both America and Europe subordinating classical education to scientific instruction. The Cornell University is such an institution. As yet it has not got fully under way, and its ultimate success or failure is problematical. We believe it will prove a triumphant success, and we have had this faith from the outset."

APRIL 1869

"The honor of the manufacturer is too often made entirely subservient to his avarice. Articles of common and daily use are made to sell, rather than to last; sham and cheapness are made to take the place of reality and worthiness; paint and putty are used to cover the lack of painstaking and patience; even labor-saving machinery is made to contribute to its quota to the revenues derived from the practice of sham. The commonest articles of household use are shams compared with those made by our fathers. Tinware will not stand scouring. In woodenware it is no better. Brooms are bound lightly with rotten twine. In the articles of furniture—common furniture for the kitchen and dining-room—it is still worse."

AUGUST 1869

"The United States may as well look the subject of Chinese labor squarely in the face, and make timely provision to absorb and utilize this new accession to our population. Some are bitterly opposed to the coming of the Chinese. This opposition is based on groundless prejudice. The policy of the Government has hitherto opened the doors of immigration to people of every race and clime. Shall we now close it on the Mongolian, and if so, why? We have heretofore spoken of the intelligence, industry, frugality and order-loving disposition of the Chinese. Since we assert that the Chinese character possesses in an eminent degree the qualities we have ever been taught to regard as the elements of citizenship, we do not see how it is possible, with any show of consistency, to attempt, either by persecution or legislation, to shut our doors against them. The Chinaman wants to work for us and we want him. Then let an end be speedily put to the disgraceful treatment he has hitherto received, a blot upon the history of the 'Golden State' which makes humanity blush. Let us welcome him with all the rest of the oppressed and suffering who now find refuge here, confident that by the process of assimilation we can absorb, and render homogeneous, the mixed races which are destined to people this continent."

NOVEMBER 1869

"There is a great deal said nowadays under the captivating title of 'Social Science,' but much of what is said and written warrants a doubt of even the existence of such a science. Still more does it warrant the doubt that those who attempt the discussion of social topics have, even admitting the existence of such a science, ever mastered the first rudiments of it. The wordy and weak discussions which have filled up the time of the 'Social Science Conventions' have not availed to fix public attention upon social evils more strongly than before they were uttered. The few suggestions made for reform and the correction of acknowledged existing evils have been of the most impracticable kind and showed most glaringly superficiality of thought in those who offered them. If there be not now, it is high time there ought to be such a thing as social science. Until some prophet arises capable of grappling with this subject from a physical and biological, as well as a political and legal, point of view, and beginning down upon hardpan shows how society may be constructed in harmony with all the conditions of pure living, regardless of creeds, conventionalities or traditions, let us not flatter ourselves that such a thing as social science exists."

SEPTEMBER 1870

"The question of what women can do and what they cannot do well is one that has been much debated of late, and it is safe to say the facts and arguments laid before the public in the course of the discussion have done much to shake the belief, once so universal, that women are adapted to doing nothing well but the domestic duties of the household. There is a great variety of occupations which women have begun to claim as fields for individual effort from which no intelligent, refined man who views things as they really are would seek to exclude them. These occupations in no way injuriously affect the qualities admired by the other sex. They may and ought to be made as remunerative to women as to men now engaged in them."

MAY 1871

"Can any one grasp the exceedingly probable fact that in 1900—only 29 years from now—the population of the U.S. will number 75,000,000? Yet, says the Evening Mail, Mr. Samuel F. Ruggles proves that this will be the case. When, therefore, the ablest, most experienced and most trust worthy statistician now living tells us that we shall have a population of 75,000,000 in 1900, the younger part of the present generation may as well consider what awaits them in their maturity and old age. Seventy-five millions of people in the U.S. implies the settlement of the entire South and West by as dense a population as that of Massachusetts; the reclamation of the arid wastes of the Great Plains by irrigation; the development of states as strong as Ohio, Indiana and Illinois along the Rocky mountains; the settlement of the Utah basin by four or five millions of agricultural and pastoral people; the development of a tier of agricultural states along our northern border from Lake Superior to the Pacific as populous and prosperous as Missouri and Minnesota; the growth of the Pacific states into commonwealths as rich and populous as New York and Pennsylvania. It means that New York will cover the whole of Manhattan Island with a population of at least two million, to say nothing of the outlying suburbs in New Jersey and across the East River."

JULY 1871

"It is said that the total number of students attending colleges in the United States is both relatively and absolutely less than it was 30 years ago. In other words, although the population of the country has greatly increased, the attendance upon colleges has either remained stationary or has actually diminished. While at one time in our history there was one person in every 1,000 in college, we now find only one person in nearly every 3,000. Our colleges would do well to consider whether the decrease in attendance may not be partly due to defects in their curriculum of instruction. Perhaps the colleges do not offer the kind of education that the times demand. It may be that a little less Latin and Greek and more of the physical sciences would be acceptable."

SEPTEMBER 1871

"The Brooklyn Union in a recent issue gives its readers an account of the extent to which opium is consumed in its city, which may be considered as somewhat astonishing. From $75 to more than $100 per annum is the cost of the opium consumption of single individuals devoted to this habit, from which the quantity they take can be estimated. It is difficult to estimate the aggregate quantity used for purposes of stimulation alone in this country or any section of it. The habit is easier to conceal than the drinking of alcoholic liquors, and statistics are hard to obtain. A country physician once remarked to us that if he could have the exclusive sale of the opium consumed in the single township where he resided, he could make his fortune without charging exaggerated prices."

OCTOBER 1871

"Those who have believed the International Working Mens' Association of small account in its influence upon industrial affairs throughout the world may learn a useful lesson from the recent struggle between labor and capital in England. About four months since, a demand was made by the workmen in the workshops on the Tyne for a reduction of one hour's labor per day without a corresponding reduction in their pay. The demand was refused, and about the first of June the workmen, numbering some 10,000, struck. A significant feature of the strike has been the united attempt made by the prominent engineering firms in England to defeat it. These sided with the Tyne firms, and raised a large fund for the purpose of importing workmen from other parts of Europe. A large number of workmen were obtained from Belgium and others were secured from the Government Arsenal in Denmark. But the influence of the International, coupled with the threats and remonstrances of the English workmen, soon overpowered that of the manufacturers. We have thus the spectacle of united capital pitted against united labor on a scale to test the relative strength of each. Prone as is the American public to refrain from recognizing and preparing for approaching emergencies, there are among us some who see that the adjustment of the relations of capital to labor will soon force itself upon public attention."

DECEMBER 1871

"In view of the enormous and increasing consumption of tobacco it has become a question of very great importance what effect upon the general standard of health is produced by it. The agitation of this subject has been increased during the past two years, and pamphlets, essays and lectures have developed in full strength the arguments for and against tobacco using. As smoking is the most popular and most powerful method by which the stimulant effect of the plant is obtained, it is principally upon this habit that the battle is waged. We have from time to time presented some of the arguments on both sides of the question, our object being to assist in arriving at truth in so important a matter; and though our confirmed taste for smoking and our natural desire to find it a harmless practice have led us to peruse with particular care all that has been said in its favor, we avow that neither reading nor experience has convinced us that the general use of tobacco is other than an unmitigated evil."

JANUARY 1872

"The habits of the present generation are such as to give rise to more refuse matter and poisonous products than those of previous ages. The fuel we use, the articles we manufacture and the waste of sewage combine to create more impurities than were known to our fore-fathers, and if it were not for the fact that science has given us reme-dies nearly in proportion to the increased evil, our population would diminish under the high-pressure system which at present prevails. It is evident that the contruction of great chimneys to carry of foul gases, together with the immense loss to agriculture, could be avoided if we applied the remedy at the outset, and that would be by using the ounce of prevention and disinfecting all animal matter by dry earth, and never allowing it to pollute our waters. Whereas our water arrangements appear to us individually a great convenience, they are collectively the source of most of our diseases and ought to be regulated. In spite of all precautions much impurity finds its way into the sewers, but the worst evil could be stayed and disinfecting rendered substantially unnecessary. The true remedy is to stop filling the sewers with matter that no power can afterward cleanse."

MAY 1872

"The most deadly physical danger in this country is the absorption of metallic poisons in water, food, medicines, washes, paints, dyes, enamels, etc., prepared and sold by the thoughtless and unprincipled. The demand of every thoughtful patriot should be that no kind of poison should be sold under any other than its proper name. The public is, and always will be, powerless to adequately protect itself against insidious poisons used in the many adulterations of the present day, and must perforce look for that protection to a government professing to guard the life and property of the citizen."

NOVEMBER 1872

"Next to our own country, there is no nation in the world that gives evidence of such rapid progress in industrial matters as Russia. Her mechanical and metallurgical interests are almost daily developing, and new means of utilizing her great resources are constantly coming into existence. A gigantic establishment has recently been founded by MM. Struve Brothers, situated near the city of Kolom, which, it is stated, rivals in magnitude the finest workshops of England or Belgium. It has been in operation but five years, and is at present engaged in the manufacture of iron bridges and rail-road freight cars. Since its foundation it has completed 3,000 cars, and since it has begun the manufacture, 79 locomotives have left its shops."

FEBRUARY 1873

"The presence of Miss Emily Faithfull in this country at the present time has revived the discussion of the question of women in society. A meeting was held a few evenings since at Steinway Hall in New York that must have given great encouragement to the advocates of the new movement. One thing is certain: The right of woman to her share of honest labor cannot be put down by ridicule or despotism. It must be met fairly and squarely, and now that it has been taken up by our most refined and gifted women we trust that the question will soon be settled to the entire satisfaction of all parties."

APRIL 1874

"Spectral analysis, confirmed by the chemical analysis of meteorites, has familiarized us with the idea that all the bodies of nature, the planets of our system as well as the suns most distant from us, are composed of the same elements and animated by the same physical forces, even to the most delicate and minute details. Since therefore these same forces act under our eyes as essential agents of life, we are naturally led to consider the conditions of organic existence on our globe as applicable to the circumstances of other species. If our earth is inhabited, why not the other orbs that fill up space, seeing that the same matter is everywhere present? The eminent French astronomer Faye, in considering this question, concludes that although the chemical elements necessary to life are largely extended throughout the universe, the requirements of temperature, water and atmosphere exclude almost all heavenly bodies, so that we are barely able to cite two planets of our system, Mars and Venus, where the conditions of life have any shadow of probability of existing, and on the only globe of which we can speak with certainty, the moon, we know them to be utterly absent."

OCTOBER 1874

"New York has a death rate such as few cities in Christendom can equal. The appalling mortality of the past summer, especially among children, has given rise to a great amount of sorrow and indignation, and not a little criticism of the medical and police authorities. That much might have been done to improve the health of the city by sanitary measures there is no doubt; the great source of disease and death in the city, however, is the tenement-house system, whereby families are massed by the hundreds in huge barracks, destitute of light, ventilation and the means of keeping clean. Only by the dispersion of the tenement-house population can the death rate be reduced to reasonable limits. There is no way by which such a desirable result can be effected humanely save by providing means for carrying the poorer working people to and from country homes more rapidly and cheaply than is possible with surface roads."

JUNE 1875

"A popular theological dogma declares that life is the grand object of creation, that the composition as well as the contour of the earth's surface has special reference to its habitability, and that all things show a ruling design to fit the world to be the home of sentient creatures, more especially of man. Strictly speaking, Science has nothing to do with such dogmas. It has no means of discovering the ultimate purposes of things, and no time to waste on their discussion. Nevertheless, it is difficult sometimes not to take an indirect interest in the claims of those who presume to decide such questions, at least so far as to notice how aptly the facts of Nature contradict their assertions. Thus in the present case it would be much easier to sustain the contrary thesis, namely that so far from having been made what it is that it might be inhabited, the earth became what it is through being inhabited. In short, life has been the means to, not the end of, the earth's development. In the light of recent discoveries, Byron's poetic extravagance, 'The dust we tread on was alive!' becomes a simple statement of observed fact."

AUGUST 1876

There is a probability that the unsightly pile of stone called the Washington monument, in the national capital, will soon be pushed forward to completion. It is now 174 feet high. The amount required to carry it up to 485 feet and surround it with a terrace 25 feet high and 200 feet in diameter, is estimated at $500,000. There is an effort on foot to raise this sum by public subscription, mainly through churches; but the patriotism of the people so far seems not to have resulted in very liberal donations. The Senate, however, has recently passed a bill declaring that Congress should assume the finishing of the work; and it will, therefore, if an appropriation be made, be paid for out of the national treasury instead of the pockets of private citizens.

JULY 1876

"We recently met with some of those axiomatic sayings of the late Dr. W. W. Hall. If on any occasion, he says, you find yourself the least bit noticeably cool, or notice the very slightest disposition to a chill running along the back, as you value health and life, begin a brisk walk instantaneously, and keep at it until perspiration begins to return. Not a summer passes but that the papers report numerous deaths from drinking ice water by overheated people. If icy cold water be taken, safety lies only in drinking slowly. Take one swallow at a time, remove the glass from the lips, and count twenty slowly before taking another. It is surprising how little water will quench the thirst when thus drank."

OCTOBER 1877

"Thanks to Stanley's pluck and energy, the well-founded belief that Livingstone's Lualaba was none other than the Congo has now been fully justified, and henceforth the Congo must rank with the three or four great rivers of the globe. It is to Africa what the Amazon is to

South America, the Mississippi to North America and the Yangtze Kiang to Asia. The German expedition under Captain Von Homeyer, which started in 1875 to explore the lower Congo to prepare the way for German colonization will probably be heard from through Stanley, when details are received of his hazardous yet successful journey. One important point in connection with the future of the Congo is already apparent: the development of the great interior basin of Africa by means of steam navigation is likely to be long delayed. The great cataracts near the Equator, no less than those near the coast, must ever present serious obstructions to the commercial development of the interior."

DECEMBER 1877

"We cannot recall in our time so gross in infringement on the rights of the people in relation to their property as is now being perpetrated in the erection of the elevated railway in New York. The effect of this outrage is to disfigure and otherwise damage several thorough-fares throughout the length of the city for the benefit of a clique of stock speculators and out-of-town land owners. We do not believe that this railway corporation has any legal right to erect its structures. When a street is opened for all kinds of public uses by compensating the land owners for the property thus taken, the government that represents the ownership of the acquired domain may authorize the erection thereupon of anything that will not interfere with such use. But even if the government can do this, the right has been most wantonly exercised, with no proper limitations and with a recklessness with regard to both public and private interests that is simply astounding. The people of New York will bitterly repent some day of this gross injustice, but the monopoly they have created, having seized its prey, will care nothing for their penitence."

JANUARY 1878

"One of the most remarkable chapters in the history of civilization is the persistent efforts of the Chinese government to prevent the importation and use of opium among its people. More than a century ago the Chinese government recognized the dangers inherent in the use of this drug and began to legislate against it. The opium was grown in India and smuggled into China by English ships, encouraged by the English government. The trade finally culminated in a war (1839), which Mr. Gladstone denounced as one calculated to cover his country with permanent disgrace. In 1857 another war was provoked by England, for the mercenary purpose of increasing her Indian treasury. The sagacity of the Chinese officials in recognizing the fatal effects of the introduction and use of opium is a lesson to our boasted civilization that we cannot afford to ignore. In this country we are just beginning to realize the necessity of some legislation to control the abuse of opium and its compounds, and no one dare deny that the same peril the Chinese discerned a century ago threatens us to-day. The opening of Parish Hall, Brooklyn, N. Y., as an asylum for opium cases alone, and the large number of patients already gathered, are significant hints of the presence of a disorder that can destroy the vitality of both community and nation."

SEPTEMBER 1882

"The word energy was first used by Young in a scientific sense, and represents a conception of recent date, being an outcome of the labors of Carnot, Mayer, Joule, Grove, Clausius, Clerk Maxwell, Thomson, Stokes, Helmholtz, Rankine and other laborers, who have accomplished for the science regarding the forces in nature what we owe to Lavoisier, Dalton, Berzelius, Liebig and others as regards chemistry. In this short word energy we find all the efforts in nature, including electricity, heat, light, chemical action and dynamics, equally represented, forming, to use Dr. Tyndall's apt expression, many 'modes of motion.' It will readily be conceived that when we have established a fixed numerical relation between these different modes of motion, we know beforehand what is the utmost result we can possibly attain in converting one form of energy into another, and to what extent our apparatus for effecting the conversion falls short of realizing it."

OCTOBER 1882

"The inaction of the committee having in charge the work of soliciting subscriptions for the foundation and pedestal for Bartholdi's Statue of Liberty has led to some impatience on the part of the French Committee of Presentation. The rumor that the latter committee were contemplating an offer of the statue to Boston has stirred the New York committee a little, but they are still debating whether to attempt to raise the money by popular subscription or to solicit a few large contributions from the wealthier citizens of the city. A member of the committee has said that various engineers have estimated the cost of a suitable base for the study at from $150,000 to $1,000,000. It is thought that an acceptable base can be made for $200,000. This sum is a mere trifle for the wealth of New York, and if the committee cannot raise it promptly, they ought in courtesy to the artist and his friends decline the proffered gift in favor of Boston or some other more deserving city."

MAY 1883

The time of our going to press slightly antedates the day of the opening of the great bridge connecting New York and Brooklyn; but our readers will be interested in knowing the intended order of proceedings.

The initial ceremonies have been appointed to take place in the Brooklyn station of the bridge on Sands Street, at 2 P.M., on Thursday, May 24, 1883.

The marshal of the day will be Major-General James Jourdan. The President of the United States and Cabinet, the Governor of the State of New York and staff, with others, will be escorted from the Fifth Avenue Hotel to the New York anchorage by the 7th Regiments, Colonel Emmons Clark commanding, and there received by the trustees and escorted to the Brooklyn anchorage, from which point the 23d Regiment, Colonel Rodney C. Ward commanding, will act as escort to the Brooklyn approach.

Seats will be reserved for the President and Cabinet, the Governor and staff, United States Senators, members of Congress, Governors of other States, members of the Legislature, the Common Councils of

New York and Brooklyn, city and country officials of New York and Brooklyn, Army and Navy, the National Guard, the Press, especially invited guests, and the employees of the bridge.

At 2 o'clock the exercises will begin at the bridge station, Hon. James S. T. Stranahan presiding.

In the evening a grand display of fireworks from the bridge takes place, and also a reception, at the Brooklyn Academy of Music, to President Arthur and Governor Cleveland.

But these exercises, however interesting to the comparatively few who can witness them, will be as nothing compared to the great popular pageant, the sight of the millions of the two cities increased by the multitudes of strangers who will march over the bridge on the opening day.

FEBRUARY 1884

"A body in England calling itself the Society for Psychical Research is addressing a series of interrogatories to the public in relation to hallucinations and dreams. An invitation is issued to all people who think they have seen ghosts or specters to state their experiences. There are some preliminary questions that ought to be asked. Has any group of presumably sane men a moral right to instigate the crazy public to formulate its 'mysterious' experiences? We know that the most disastrous consequences sometimes ensue to weak brains from dwelling too intently on fixed-ideas. Such things exist only in the imagination of the persons who are subject to them."

SEPTEMBER 1885

"By a recent enactment of the Pennsylvania legislature boys under fourteen years of age, and all women and girls, are prohibited from being employed in the coal mines of that State. It is estimated that the law covers nearly one-half of the whole number of slate pickers in the mines, at which boys are sometimes employed when only six years of age, and that it also includes a good proportion of the mule drivers and door tenders. It has required many years of agitation to get the law passed, and its enforcement now is causing no little excitement in the mining regions. Yet society undoubtedly owes it to itself to see that these little ones are at school instead of being thus early predestined to a life of ignorance."

APRIL 1888

"The number of women of our country who have under-taken and are carrying on business enterprises successfully are not a few, and they increase every year."

AUGUST 1888

"The importation of fire crackers this year will amount to 300,000 boxes, an increase of 100,000 boxes over last year's importation. It is a little curious that the scientific knowledge and inventive genius of this country have proved inadequate for the successful manufacture of these explosives. All attempts to produce them in this country, so as to compete with the imported article, have failed. They are made in China and Japan, and the importation of the last week in June was 14,415 boxes, valued at $34,255. What a large sum to be thrown away on such trash!"

APRIL 1889

"On the question of the right to photograph there is as yet very little judicial opinion. Clearly, every one who sees fit may make pictures of natural scenery. There are objects, however, which cannot be photographed without the consent of the owner. Furthermore, suppose the rejected suitor of the fair Amanda, by means of a detective camera, succeeds in obtaining a picture of his adored one in the very act of being kissed by his hated rival. Would he be allowed to exhibit such a picture? We think that no photograph exposing others to scorn, disgrace, humiliation, or contempt would be tolerated."

AUGUST 1889

"Observation in public places gives satisfactory evidence that the use of cigarettes is rapidly on the decline. Whether this is because of the stringent laws passed in many of the States against selling them to minors, or because smokers have come to their senses and have taken warning from their own experience and the unanimous condemnation of smoking cigarettes by the medical profession, we know not."

AUGUST 1890

"A wretch named Kemmler, whose crime had been the atrocious murder of a woman, was appointed to be the first to suffer electrical death. The doomed man was strapped to a stout chair, electrodes were placed so as to make contact with top of head and base of spine, an alternating electrical current from a powerful Westinghouse generator was joined, a switch was moved, and the criminal was struck dead—instantly killed by lightning. The execution of a criminal, whether by the guillotine, the garrote, the gallows, the gun, or the dynamo, is a ghastly business; and it is not surprising that the sensational newspapers, aided by the electrical opponents of the law, should have made the most of such an occasion to fill their columns with revolting details."

JANUARY 1893

"A bill has recently been presented in Congress requiring the Secretary of the Treasury to provide for the calling in of all ragged, worn, and soiled paper money, new bills to be furnished in place of the old and unclean notes. It is surprising that some such action has not long ago been taken, for not a little of the paper money daily passing from hand to hand has become extremely repulsive in appearance, and is ever suggestive of disease-spreading power."

APRIL 1895

"An ideal school room should provide fifteen square feet of floor space for each pupil and a supply of 200 cubic feet of air per minute for every person in the room. Such provisions would ensure the free movement of every child and a wholesome amount of air. In France, the perfect school room, it is thought, should have a window area equal to one-fourth the floor space. It is also thought best to have individual seats and desks for the pupils."

OCTOBER 1896

"Cycling, which was yesterday the fad of the few, is today the pastime of the many. Unfortunately, this progress has been attended with numberless casualities. One temptation to many cyclists is to see how speedily they can sacrifice their lives on hilly ground. The moment the brow of a hill is reached the reckless cyclist seems impelled to take his feet from the pedals and to allow the machine to descend with all the rapidity which gravity gives it. A good brake affixed to the back wheel would considerably reduce the number of accidents from this cause; but, unfortunately, there is an idea that a brake adds an inconvenient weight to the machine."

MARCH 1897

"In a recent lecture before the American Geographical Society, Mr. Heli Chatelain made some very startling statements regarding the extent and horrors of the slave trade in Africa. Let no one suppose that the slave trade in Africa is a thing of the past. In this great continent, which the European powers have recently partitioned among themselves, it still reigns supreme. 'The open sore of the world,' as Livingstone termed the internal and truly infernal slave trade of Africa, is still running as offensively as ever. Among 200,000,000 Africans, 50,000,000 are slaves. In the islands of Zanzibar and Pemba alone, which are entirely governed by Great Britain, 260,000 are held in bondage. For each slave that reaches his final destination, eight or nine are said to perish during the journey, so that the supply of 7,000 slaves annually smuggled into Zanzibar represents the murdering of some 60,000."

AUGUST 1897

"The announcement of the return of two steamers from the Alaskan gold fields along the Klondike River last month, with a small party of miners on board who carried about a million and a half dollars in gold between them, has gone through the world like an electric shock. The news is expected to set off a 'gold fever' comparable only to the wild excitement of the California discoveries in 1849. Already the 'rush' has begun, in spite of the warnings of the miners who have just come out of the country, and the detailed account by the press of the inhospitable and inaccessible nature of the placer districts."

JUNE 1898

"The growing danger of slaughter houses as a factor in spreading infectious disease is at last being appreciated. Ch. Wardell Stiles, Ph.D., in a paper published in 1896, says, 'When the offal of a trichinous hog is fed to hogs which are raised upon the grounds, the latter cannot escape infection with trichinae. Every slaughter house is a center of disease for the surrounding country, spreading trichinosis, echinococcus disease, gid, wireworm, and other troubles caused by animal parasites, and tuberculosis, hog cholera, swine plague and other bacterial diseases.' He recommends: (a) Offal feeding should be abolished; (b) drainage should be improved; (c) rats should be destroyed; and (d) dogs should be excluded from the slaughter houses."

JULY 1898

"Of the 298 classes of objects of fear to which 1,707 persons confessed, thunder and lightning lead all the rest. But is there any factual justification for this fear? We believe there is not. As proof we may cite statistics of the United States Weather Bureau. For the years

1890-1893 the deaths from lightning numbered an average of 196 a year. Indeed if one can go by statistics, the risk of meeting death by a horse kick in New York is over 50 per cent greater than that of death by lightning."

AUGUST 1898

"It has been considered a quite sufficient educational training for the young to cram and overload their brains with a quantity of matter difficult to digest, and of little use in after life. Numbers of delicate, highly strung children have broken down under the strain, and the dreary daily grind has developed many of the nervous diseases to which the present generation is so peculiarly susceptible. However, Americans are becoming alive to the pernicious effects of developing the mind at the expense of the body, and in the ten years since German gymnastics were introduced, physical training holds a place in the curriculum of most larger schools."

FEBRUARY 1899

"One year ago a company put thirteen horseless electric cabs for hire on the streets of New York. Today the same company operates one hundred cabs. With even the partial exit of the horse will disappear to a great extent the dust and mud and noise on the cobblestone pavements, and it will benefit the public health to an almost incalculable degree. The noise and clatter which makes conversation almost impossible on many streets of New York at the present time will be done away with, for horseless vehicles of all kinds are always noiseless or nearly so. Specialists have expressed an opinion that the nervous diseases which exist in the city are aggravated, if not caused, in many cases, by noises incident to a great city's traffic. The bells of the new vehicles will of course be somewhat annoying at first."

JUNE 1901

"Science and art are becoming more and more the mere hand-maidens of industrialism. Our greatest scientific men are devoting their energies, not to pure science, not to their noble profession in its abstract or elementary form, but to those applications of it which result in some new economy of the world's work and in the formation of more immense stock companies, with bonds and common and preferred shares, dividends, and all the paraphernalia of modern financial operations on a big scale. The men who love science for science's sake are giving way to the Edisons, Teslas, Triplers, Pupins, Marconis, those wizards who by day and by night seek to wrest from nature some new and commercially profitable service to mankind. The number of patents taken out at Washington steadily increases, notwithstanding the predictions made not long ago that American inventiveness had reached its high tide. This is the age of materialism and of Mammon, sure enough."

AUGUST 1902

"In the United States the use of narcotic plants is confined mainly to tobacco smokers, but it is interesting to note that the use of Indian hemp is spreading throughout the Southwest. The effect of this drug is well known from accounts published in the daily press and elsewhere. The common Mexican name of the plant is *mariguana*. It is not uncommonly asserted by Mexicans that sometimes a single dose of hemp will cause long-lasting insanity. Van Hasselt, a Dutch authority on poisonous plants, also asserts that a single dose of this drug may cause mania for months but the best pharmacologists are agreed that such might be the case only when the person affected is already badly diseased by the use of drugs or otherwise. The Department of Agriculture is at present engaged in an investigation of the curious behavior of these weeds."

SEPTEMBER 1902

"At the annual meeting of the British Association for the Advancement of Science, Professor Dewar made a stirring appeal for the improvement of the national system of scientific education. As an instance of the importance of science to a country, he pointed out that the German chemical industries, which have grown up during the last

70 years, are worth £50,000,000 annually. Curiously enough, these chemical industries are founded on basic discoveries made by English scientists. 'It is in an abundance of men of ordinary plodding ability, thoroughly trained and methodically directed, that Germany at present has so commanding an advantage. It is the failure of our schools to turn out, and of the manufacturers to demand, men of this kind, which explains our loss of some valuable industries and our precarious hold on others. The really appalling thing is not that the Germans have seized this or that industry, or even that they may have seized a dozen industries. It is that the German population has reached a point in general training and specialized equipment which will take us two generations of hard and intelligently directed educational work to attain: it is that Germany possesses a national weapon of precision, which must give her an enormous advantage in every contest depending upon disciplined and methodized intellect.'"

FEBRUARY 1903

"Some time ago the British Board of Trade was able to announce that during a period of twelve months not a single passenger had been killed on the rail-roads of Great Britain. Since then another three months has passed without a fatality. Here, in the United States, our railroads have killed 77 passengers in fifteen days! We recently presented this comparison of railroad fatalities to the chief engineer of one of the leading rail-roads. In his prompt reply he put his hand at once on the weak spot: 'The different results are to be explained by a difference in national temperament—here, we take chances.'"

SEPTEMBER 1906

"Frederick Soddy discusses the unknown store of energy contained in the chemical elements, and the prospect of making it available. None of the material changes dealt with in chemistry is very profound. Radioactivity has brought more fundamental changes into view. The large internal energy exhibited by the radium atom cannot be regarded as peculiar to radium. Uranium was known long before its radioactivity was discovered, and represents another perfectly normal chemical element. Yet uranium, since it produces radium with evolution of energy, must possess all the internal energy of radium and more. It is probable that the elements as a class all possess great internal energy, and that their characteristics of stability and permanence, and the failure of all attempts to change them by artificial means, are due to the existence of this internal energy. The forces at our disposal compared to those which are exhibited when an atom suffers change, are of a different and lower order of magnitude, and it is not to be expected therefore that transmutation will become possible until we can control more powerful agencies than are at present available. Suppose that a way were known in which the element uranium, for example, which disintegrates to the extent of a thousand-millionth part annually, could be made completely to disintegrate in the course of a year. From one gramme of the element more than a thousand million calories could be evolved, and this, if it could be converted into electrical energy, would be equivalent to more than 1,000 kilowatt-hours, and would suffice to keep a 32-candle-power lamp burning continuously throughout the year. By the expenditure of about one ton yearly of uranium, costing less than £1,000, more energy would be derived than is supplied by all the electric supply stations of London put together."

AUGUST 1908

"It is not often that a measure of such startling character as the Daylight Saving Bill is introduced into the English House of Commons. It is proposed, during part of the spring and autumn and the whole of the summer, to advance the clocks throughout the country, moving the working day forward, with a view to including within the working hours a longer stretch of daylight. The advantage of a long daylight evening for such sports as yachting, rowing, golf, tennis and automobiling is indisputable."

APRIL 1909

"To the Editor of SCIENTIFIC AMERICAN: An airship is either a 'heavier-than-air machine' or a 'lighter-than-air machine.' But these are very clumsy names. Why not call the former a 'pondro,' and the latter a 'levitar'? These words, I think, are sufficiently 'regular' in derivation to justify themselves, and they are not awkward. Ambrose Bierce, Washington, D. C."

MARCH 1910

"Every day some four or five million persons attend the 13,000 moving-picture shows of the United States. The pictures which flicker on the screen before the spectators are projected by means of apparatus the basic patents of which were taken out by Thomas A. Edison. These four or five million persons, therefore, unwittingly pay royalty to Mr. Edison whenever they hand in their nickels or dimes at the box office. It is stated that about 1,440,000 feet of film are made by the members of the Motion Pictures Patents Company. On this production a royalty of half a cent per foot is paid to Mr. Edison, so that his revenue from the moving-picture-loving public amounts to $7,200 per week."

JANUARY 1911

"Had the incident occurred a quarter of a century ago we could have understood better the discussion which has been provoked in that august body, the Academy of Sciences, Paris, by the candidature of Madame Curie for membership. Respect for custom and tradition is an admirable attitude if it be judiciously tempered by due considerations of time, place and personality, but we cannot help feeling that in this advanced age, in such a center of enlightenment as Paris, and where a scientist of such brilliant performance as Madame Curie is concerned, this discussion as to whether she is eligible for admission to the Academy of Sciences is altogether deplorable. When science comes to the matter of bestowing its rewards it should be blind to the mere accident of sex; one does not have to be an enthusiast on the subject of the extension of the rights and privileges of her sex to feel that here is a woman who, by her brilliant achievement, has won the right to take her place with her compeers in the Academy, or any similar institution devoted to the furthering of science."

JULY 1911

Capt. Amundsen, who left Norway ostensibly for the Arctic regions, where he proposed to make a five-year drift across the Polar Sea, recently turned up, to the surprise of the scientific world, in the Antarctic. His change of plans has now been explained by Dr. Nansen in a letter to the *London Times*. It appears that he wrote to Nansen from Madeira that, owing to the diminished popular interest in the North Pole since the successful result of Peary's last expedition, he felt convinced that he would not be able to raise sufficient money for the proposed voyage in the Arctic. He therefore decided upon the more popular and less expensive plan of vying with Scott, Filchner and the others in a dash for the South Pole. A press dispatch, dated June 17, states that Pedro Christophersen, a Norwegian in business in Argentina, has agreed to finance Amundsen's expedition to the extent of $830,000."

AUGUST 1911

"Is there any reason why we Americans, admitted to be the most ingenious people in the world, should suffer as our men do in the summer season because of our unreasonable style of dress? Notice a group of men and women in summer: while the women are attired in their cotton or linen garments, the men will have on coats as warm as those worn by many women in the depth of winter. No one has as yet been able to devise or suggest any acceptable summer dress for men, who continue summer after summer to swelter in warm woolen garments. Is this not a fit subject for invention?"

APRIL 1912

"On Sunday, April 14, the largest and supposedly the safest steamship afloat, while steaming on her proper course, on a clear, starlit night, struck an iceberg and within a few hours sank, carrying down with her more than 1,600 souls. The loss of the *Titanic* has brought home to the public at large the fact that, in spite of the improvements in ship design and construction, there is not a vessel afloat that is unsinkable by one or other of the accidents to which ocean travel is liable. And this is not to say that a great advance toward the unsinkable ship has not been made. If then, the modern ocean liner is not unsinkable, dictates of common prudence and humanity demand that it should carry a sufficient number of lifeboats to accommodate every soul on board. Under an international agreement our government accepts the certificate of inspection of foreign countries; and if the Board of Supervising Inspectors finds that the foreign ship carries the number of boats called for by the certificate, she is permitted to sail. Had the *Titanic* carried the American flag, she would have had to provide space in her lifeboats for 2,412 passengers and crew. As it was, the maximum provision in the lifeboats that the Titanic carried was about 1,000! In the presence of this stupefying disaster, we enter a plea for the exercise by Congress of a calm and judicial spirit in all legislative action that may be taken. Evidently the matter is one for joint international action."

OCTOBER 1914

"Surely one of the most appalling things about the present war is the fact that some of the most brilliant young men in Europe, men who cannot be replaced, have been ruthlessly butchered. These men belong to wider issues. Science is more important than the preservation of any one country's independence. An effective remedy for cancer is worth a colony. It is merely a question of economics. As a matter of probability, the loss of these lives entails an inconceivably greater loss than any gain their presence in the fighting line is likely to bestow."

MAY 1916

"The war has been a great revealer of national character, and the revelation has been full of the unexpected and surprising. Those of us who appreciated the genius of the German people for organization and efficiency, and admired that strong logical bent which enabled them to move with such directness to their great industrial and commercial accomplishments, have been dumfounded by the total lack of moral and ethical qualities as revealed in the gospel of might and frightfulness which the Germans have preached and practiced throughout the war. As evidence of this, consider the violent recrudescence of the murderous raids of the Zeppelins, whose victims are almost entirely unarmed non-combatants, at the very time when the German government professes to be endeavoring to meet the humanitarian views of President Wilson on the subject of submarine warfare. That Germany should increase her activities in the one field at the very time when she is supposed to be looking for some reasonable basis on which to diminish her activities in the other field is the latest of those many amazing contradictions that have made the civilized world ask over and over again. 'What manner of people is this?' One of two alternatives is certain. Germany, in this wholesale running amuck among non-combatants, not only of the belligerents but also of the neutral powers, is doing so either with cold-blooded but clear-headed and deliberate intent, or she is proving on a most tragic scale, that brooding over fancied wrongs and too-long-imagined plots and persecutions may produce insanity in the nation even as in the individual."

FEBRUARY 1917

"The President of the United States appeared some days ago before the Senate and made a most notable and unusual address which has attracted the attention of all civilized countries and has brought both criticism and praise for its many suggestions and proposals. His principal thought has apparently been the organization among nations of a league to enforce peace—by some styled a dream, by others something that is practicable. At any rate, a few extracts from his speech will show the at least startling nature of his statements. In one place he remarks, 'No peace can last or ought to last which does not recognize and accept the principle that governments derive all their just powers from the consent of the governed and that no right anywhere exists to hand peoples about from sovereignty to sovereignty, as if they were property.' What would disturb our people most in a league with other nations would be the 'entangling alliances' against which Washington cautioned his fellow countrymen: the Monroe Doctrine would apparently also be set aside. But under this head the President makes a strong argument: 'I am proposing, as it were, that the nations should adopt the doctrine of President Monroe as the doctrine of the world.'"

JULY 1917

"A few days ago a correspondent of the *Daily Mail* resuscitated a well-known quotation from George Gissing's *Private Papers of Henry Ryecroft* in order to associate science with the horrors of the present war. We have on several occasions pointed out that it is merely pandering to popular prejudice to make science responsible for German barbarity or for the use of its discoveries in destructive warfare. Chlorine was used as a bleaching agent for much more than a century before the Germans first employed it as a poison gas, and so it is with other scientific knowledge—it can be made a blessing or a means of debasement. The terrible sacrifice of human life we are now witnessing is a consequence of the fact that the teaching of moral responsibility has not kept pace with the progress of science. As in medieval times all new knowledge was regarded as of diabolic origin, so even now the popular mind is ever ready to accept such views of the influence of science as are expressed in Gissing's work. The pity of it is that the public press does nothing to dispel illusions of this kind by urging that what is wanted is not less scientific knowledge but a higher sense of human responsibility in the use of the forces discovered."

DECEMBER 1918

"When the naval section of the Peace Conference settles down to its work, we suggest that one of the first, if not the very first, subjects to be considered should be the placing of a ban upon the construction and use of the submarine. This should be done, first, on the ground that as a lawful weapon of war it has failed of its purpose; second, that as an instrument of piracy and a menace to the freedom and safety of the seas it has proved to be a weapon of frightful potency. Therefore let Civilization put the thing under the ban. That can be done at once and for all time."

JUNE 1919

"Perhaps the greatest significance of Major Hoke's recent invention of a precision block gage accurate within millionths of an inch is that it has forced the adoption by the Bureau of Standards of the wavelength as the measuring element. At present the standard yard is defined as the length between two marks on a certain metal rod at a certain temperature. This is not ideal, for several reasons. In the first place, why should one have to go to Washington or Paris to get a standard length? Would it not be more rational to define the yard, the meter, etc., in terms of something that could be reproduced independently anywhere? Would it not be a good idea at the same time to seek a standard that had no relation to temperature? It surely would, and the whole procedure in the Hoke case fairly thrusts such a standard upon our attention. Why should we not define the yard and the meter as so many wave-lengths of such a particular light? For testing Hoke gages cadmium light has been found to possess several notable advantages. Major Hoke looks forward to the day when the standard of length will be defined as so many wave-lengths of such-and-such a light-cadmium perhaps, or something else if anything better can be found."

APRIL 1920

"Popular interest in Dr. Goddard's rocket for reaching high altitudes was excited by the claim that this projectile could actually be made to travel to the moon and there flash a signal that would show that it had completed its journey. There is something romantic in the thought of crossing the intervening hundreds of thousands of miles to the faithful satellite that is our closest companion in the infinite reaches of space. To be sure, there would be little, if any, astronomical value in such an accomplishment. It would serve merely as a demonstration of the power of man to overcome seemingly insurmountable handicaps."

AUGUST 1921

"In the ruthless destruction of our forests and the extravagant and wasteful methods by which we are using up the natural resources of

America we have been following a policy that has been truly described as one in which it is a case of 'every man for himself and the Devil take the hindmost.' The most discouraging fact about the whole situation is that, in spite of endless warnings and the carefully prepared governmental statistics showing the rapid depletion of our resources, particularly of our forests, nobody seems to be very much disturbed and the movement to correct this abuse is apparently making very small headway."

DECEMBER 1922

"Radio broadcasting in its present form was built up overnight. Due to its complex nature, the lack of precedent or parallel to go by and other deterrent factors, it was to be expected that this hastily and poorly erected institution should be inherently weak and unsatisfactory to the radio industry and the public alike. It should be taken down to its very foundations and reconstructed along safe and sane lines. There are too many poor broadcasting stations and not enough good ones. The latest figures indicate that on October 5 there were 546 broadcasting stations in the U.S., very spottily distributed throughout the country, and up until recently all these stations have been operating on one and the same wavelength, 360 meters, giving rise to endless confusion. Broadcasting caught our legislators unprepared. The present radio law was designed before radio telephone existed in everyday form. There is a bill in the House and Senate at the present time which, if passed, will give to the Secretary of Commerce regulatory powers, and this bill should be brought up for public hearing without delay. The proper radio laws, giving several bands of wavelengths for different classes of stations, might do much to increase the possible number of broadcasting transmitters with a minimum of interference."

SEPTEMBER 1924

"Not so long ago an orator had to exert himself to the utmost to have his voice reach one or two thousand persons in a large auditorium. Today that same orator, without raising his voice above a conversational tone, can make his voice heard by tens of thousands of people indoors or outdoors. This is achieved by putting kilowatts of electrical energy into the speaker's voice and having numerous loudspeakers face the listeners. In both the Republican and the Democratic national conventions comment was frequently made on the careful attention paid to the speakers. The reason, not always appreciated, was that the entire audience, even in the farthest seats, could hear plainly. Moreover, voice amplification more or less levels off the speakers' ability to shout, and hence indirectly puts a premium on what is said rather than on how loudly it is said."

MARCH 1925

"Great changes in human affairs take place inconspicuously. The substitution of bronze for flint, of iron for bronze, of the printing press for the scribe and of mechanical power for human labor—each of those events occurred so gradually that probably even those directly concerned hardly realized what was going on or appreciated its significance. A case in point is a cultural change now in progress that promises to be as profoundly revolutionary as any that have preceded it. This change is the gradual abandonment of man's most ancient tool—fire. The first effective step toward a fireless future was the substitution of the electric lamp for a flame for illumination. Next came the use of the electric motor in the place of numberless small steam engines and their necessary boilers and fire boxes. The next step, and the one in which the electrical industry is at present particularly interested, is the substitution of electricity for fire in producing heat for industrial purposes."

JULY 1925

"To the followers of Pythagoras the world and its phenomena were all illusion. Centuries later the Egyptian mystic Plotinus taught the same doctrine, that the external world is a mere phantom, and the mystical schools of Christianity took it up in turn. In every age the mystically inclined have delighted in dreaming that everything is a dream, the mere visible reflection of an invisible reality. In truth the delusion lies in the mind of the mystic, not in the things seen. The alleged untrustworthiness of our senses we flatly deny. We frequently misinterpret the messages they bring, it is true, but that is no fault of the senses. The interpretation of sense impressions is something to be learned; we never learn it fully; we are liable to blunder through all our days, but that gives us no right to call our senses liars. It is our judgement, not the sense of sight, that is occasionally deceived. We not only wrong our honest senses but also lose our grip upon this most substantial world when we let mistaken metaphysics persuade us to doubt the testimony they bear."

JUNE 1926

"The Royal Society of London has been asked by one or more of its members to consider the advisability of requesting Sir Oliver Lodge to resign from the Society on the ground that his views on spiritualism are prejudicial to the interests of the Society. It is incredible that this great institution should assume the right of censorship on all expression of personal views by its members on general scientific subjects. On the other hand, since Sir Oliver, because of his wonderful facility in writing down to the understanding of the general public, has such a widely extended vogue, we think that he should be more careful to make it known that when he writes as a so-called spiritualist, he is giving merely his personal views, and that these views are not to be taken as an expression of the attitude and belief of the great Society to which he belongs."

FEBRUARY 1928

"How far in this age of frankness can a journal such as SCIENTIFIC AMERICAN go in the discussion of sex? We recall that in our time some of the youth obtained their information about sex from their parents. Most of us, however, received it from older youths and from books that in some way implanted in our young minds the falsehood that sex was bad. We therefore make no apology for mentioning here a series of pamphlets we sincerely believe the majority of our readers should know about: the recent publications of the National Committee for Mental Hygiene. These several pamphlets will inevitably irritate many, for the truth always irritates. They will shock a few of the tender-minded, and they will doubtless confound utterly some of the misbeliefs even of the scientifically minded. What are the 'average' sex practices of the human race? We had thought we knew, yet it appears we did not. We have often mistaken general impressions and traditions for fact. We must liquidate our present beliefs and remold them on a more reasonable basis. In short, if what we have thought to be abnormal in sex life now turns out to have been more the rule, must we not now revise the very criterion of what actually constitutes an average normal sex life?"

MARCH 1928

"An interesting set of circumstances surrounding the use of a trademark has recently been brought to light by the unsuccessful attempt of the B. F. Goodrich Company to prevent the Closgard Wardrobe bags. The well-known hookless Fasteners are made by the Hookless Fastener Company and sold to various manufacturers. The Goodrich Company buys the fasteners and places them on boots and shoes made of rubber and fabric. In order to popularize this footwear the Goodrich Company coined the term 'zipper' and registered it as a trademark. Along comes the Closgard Company prepared to reap where it has not sown. It applies for registration of the name 'zipper' for wardrobe bags. The Hookless Fastener Company cannot protest: it makes the fastener but has nothing to do with the trademark. The Goodrich Company protests in vain, for although it originated the trademark and made it valuable, it has applied it only to footwear."

MARCH 1928

"Evolution, eugenics and the breeding of new varieties of crop plants and domestic animals—all these are affected by the work of Professor H. J. Muller of the University of Texas. Stated in three sentences, this is what Professor Muller's experiments signify: Evolutionary changes, or mutations, can be produced 150 times as fast by the use of X rays as they can by the ordinary processes of nature. This means that man may someday force the production of new and desirable plant and animal varieties far more rapidly than he has hitherto been able to get them. But X rays affect the human's heredity cells too, and the reckless exposure of these cells to long and heavy doses of the rays is apt to inflict fearful penalties on our unborn grandchildren."

AUGUST 1928

"Evidently the 'talking movie' is with us to stay. A research branch of the Western Electric Company has recently announced that contracts have been signed indicating that the major motion-picture producers in the country will adopt one form or another of talking movie. But deliver us from the speaking voice of some of the present-day motion-picture stars! What a sad disillusionment it is to see the sweet face of a popular star, only to hear when she speaks the coarse, harsh voice of the forewoman of the local hat factory. Will the advent of the talking movie mean that a new set of stars will rise in the firmament of moviedom? Or will 'ghost talkers' spring up in the industry?"

OCTOBER 1928

"It is astonishing to discover in this boasted 'age of reason' how many millions of otherwise sane, level-headed, intelligent people still dwell, after a fashion, in the very midst of the Dark Ages. We refer to the recent craze for astrology, the 'science' of the stars in their control over the destiny of human individuals. One need only visit the corner bookstore to learn that there is at present an enormous sale of books on the subject: they have broken into the ranks of the very best sellers. Astrology was the parent of astronomy, but when astronomy came of age, it parted company with its parent, ashamed of its origin. Do the good souls who believe the stars exert some influence over our lives know what stars are? Let them buy a dollar book on astronomy, the offspring of their beloved occult study of subtle influences, and find out what sort of thing is the universe in which they live."

JULY 1929

"Everybody knows the end products of scientific investigation because everybody uses them, but few know the stages by which these end products have developed. Many see only what is on top. They accept, for example, the blessings of our knowledge of germs and germicides without knowing how Pasteur discovered germs as the chief source of disease; they use a large variety of chemical compounds without knowing the meaning of chemistry; they admire skyscrapers, great bridges and other complicated structures not knowing how Galileo laid the firm foundation of the laws of forces without which these structures could not be safely erected. There are millions of otherwise intelligent people whose outlook on the world is essentially that of the Middle Ages. A large number confuse science with pseudo-science and are easy game for all kinds of quackery purporting to cure bodily ills. The people of a sovereign state attempt to settle by referendum the question of whether man is a product of organic evolution; for them to attempt a popular verdict on the Einstein theory would be no less absurd."

NOVEMBER 1929

"If 10 economists were asked to give the cause of the great bull market in securities that has raged almost unabated for the past five years, we would no doubt obtain various answers. The layman would be quick to say that it was due to prosperity. That would be a true but superficial answer. Any theory advanced, if at all scholarly, would have to include an analysis of the influence of research and invention upon the trend of security prices. The busy tickers of Wall Street today reflect not only good management and good prospects but also the possibilities of new products from research laboratories. The modern investor, if he is discreet and cautious in the purchase of securities, not only scrutinizes the board of directors but also extends his investigation to the research department of the corporation whose securities he anticipates buying. The real wealth of the modern corporation and the potentialities for enhancing its earnings are created in the laboratory."

JULY 1932

"Archie M. Palmer, associate secretary of the Association of American Colleges, has described the means in which various universities throughout the country have met the question of what they shall do with valuable and patentable discoveries made by faculty members. Perhaps the outstanding patent now held by any university is that by which the University of Toronto controls the manufacture of insulin. 'So successful,' writes Mr. Palmer, 'has been the administration of the patent rights with respect to this product that the income of the university for one year was $500,000.' In treating the legal problems arising from the financing of research by industry the method employed by the University of Michigan is one workable solution. 'There the Department of Engineering Research performs considerable experimental and research work in its laboratories under contractual arrangements with industrial concerns. However, the University reserves the right to publish for the benefit of science such results as are in the nature of fundamental principles.'"

MAY 1933

"When Lindbergh flew to Japan, some newspaper comments had it that he was not flying the shortest route from the West Coast to Japan. He flew across Canada and via Alaska and the Aleutian Islands. Lindbergh was perfectly right. Our notions of distances are often based on maps in which distances are distorted. If we look down on a globe from above the North Pole, we see that the shortest way from the United States to China is over the Arctic region. The shortest way from San Francisco to England is not via New York and the North Atlantic but across Canada, Greenland, Iceland and the Faroe Islands. We must revise our notions of the shortest routes to other continents. When the United States is connected by regular airlines to

China, Japan, Siberia and Europe, it may well be that these airlines will all go near the Pole. Mankind is never daunted by difficulties if a worthwhile objective is to be obtained, and plane designers, inventors and airplane flyers and operators may eventually transform the wild regions of the North into a busy sea of aerial activity."

DECEMBER 1933

"The doctrine of Nordic superiority is an off shoot of Aryanism, the chief exponent of which was Count Joseph Arthur de Gobineau, a French aristocrat who died in 1882. Gobineau maintained that one race alone, the Aryans, has been creator and sustainer of all that is great and good in civilization. The idea of an Aryan race was based on the discoveries of similarities in the languages of the Indo-European group, which led to the theory that all these languages were derived from a common stem, the Aryan language. Gobineau and his disciples assumed that the existence of an Aryan language implies existence also of an Aryan race. Having created this mythical race, they attributed to it all virtue and excellence, and they saw in it the source of every great civilization of antiquity and of modern times. The Nordics were represented as descendants of the original Aryans who settled in northern Europe and from whom in turn came the Teutonic and Anglo-Saxon peoples. In spite of all efforts no one has ever been able to produce the slightest bit of real evidence that any such race ever existed. There is no necessary relation between language and race, and the very use of the term 'Aryan' in a racial sense—as the Germans are using it today—has no justification whatever."

FEBRUARY 1936

"It is commonly granted that the motion picture is important not only for its pervasive social effect but also because it is the only new artform of modern times. Unfortunately films themselves are singularly evanescent. Generally speaking, a film two years old is a film that will not be seen again; the situation is comparable to that which would be created in the world of literature if only those books published within the past 12 months were available. In order to remedy this situation the Museum of Modern Art Film Library has been established for the purpose of collecting and preserving outstanding motion pictures of all types and of making them available to colleges and museums, thus to render possible for the first time a considered study of the film as art."

AUGUST 1936

"It is fashionable in certain quarters to attack the modern streamlined car as unsafe. The commonest criticism is that these efficient cars are too highly powered and therefore tempt drivers to excessive speeds. Another criticism is that streamlining and lower passenger seating decrease visibility. It is also claimed that modern cars have poor roadability. High steering ratios are said to produce sluggish control. Low ground clearances are held to be a source of danger to the transmission and exhaust systems. Poor ventilation and exhaust gas fumes are stated to be the direct cause of frequent accidents. A careful statistical study made by the Travelers Insurance Company seems to refute these views, and to place greater blame on the driving public than on the design or condition of automobiles in service."

OCTOBER 1936

"Every now and then someone recalls the widespread early predictions that radio would prove of tremendous educational value. Instead of 'educational', let us use 'enlightenment.' On such a basis, radio broadcasting has done so much good for all the people that we may overlook its more obvious faults. Amazement has of late often been expressed at the conversation of the man-in-the-street on economics and sociology, politics and international affairs, war and peace and the pros and cons of almost every conceivable subject. Over the radio he has heard the voices of political candidates, of things and dictators, of opera stars and commentators on everything under the sun. He has heard leaders in every walk of life discuss and argue the subjects closest to their hearts. There is your education! It is enlightenment and culture."

AUGUST 1937

"Believing that instruction in the theory of the rules of the road and actual practice in driving a car have a place in the curriculum of the present-day high school as a means of promoting automobile safety, the American Automobile Association has sponsored a driver training program. It already has met with great success in ten high schools."

DECEMBER 1937

"Thanks to the common sense and co-operativeness of 11 governments representing both the economic interests of the whaling industry and the broad interest of science in the preservation of our remaining fauna, that valuable and ever-fascinating aquatic mammal the whale, by far the hugest bulk of animate flesh this old earth has known at any period, is now probably saved. Game laws are now to be applied to whales. These game laws are to go into effect in the Antarctic regions on December 8, and whaling will continue only until March 7."

JANUARY 1938

"Increased use of tar on roads throughout the country is causing the water supplies of hundreds of cities to take on objectionable tastes and odors, reports the American Institute of Sanitation. Road tar contains small amounts of phenolic chemicals, which are leached out by the rain and carried along to the lakes, rivers and reservoirs from which cities obtain their water supplies. The chemicals are usually present in very small amounts and ordinarily are unnoticeable to the taste. But when the water is chlorinated, the phenolic substances are turned into pungent compounds having a pronounced medicinal taste."

APRIL 1938

"In the light of its astounding capacities, in the light of its protein nature, and viewed against the background of all distinctly non-living arrangements of atoms and all living organisms, the virus molecule must for all time appear to be a transition form between non-life and full-flowering life—but a form vividly more alive than not."

MARCH 1939

"The problem of stream pollution is of direct concern to those who gain a livelihood in the fisheries industries, to consumers of shellfish and fin fish, and to those who at intervals fortify their health and strength with recreational fishing. To anyone who has noted the increased demands of the public in recent years for the effective cleaning of streams and coastal waters, it must be apparent that the time is rapidly approaching when an intelligent and comprehensive plan for dealing with the problem must be evolved. If industry and government are co-operative in finding a solution, it will be easier, less costly and less disruptive than if a thoroughly aroused public, irked by delays and subterfuge, finally insists on immediate, drastic and precipitate action."

DECEMBER 1940

"History records several cases in which entire nations have within a few years developed wide-spread alcoholism. In all such cases the cause has been the introduction into popular consumption of high proof spirits with low taxes, or none. This occurred in England in the reign of Queen Anne. England was at war with France, and allied with Holland. So French wines were so far as possible excluded, and 'Hollands,' or gin, was favored as a sign of patriotism. As a result, habitual drunkenness became appallingly general."

MARCH 1942

"Bischoff, one of the leading anatomists of Europe, thrived some 70 years ago. He carefully measured brain weights, and after many years' accumulation of much data he observed that the average weight of a man's brain was 1350 grams, that of a woman only 1250 grams. This at once, he argued, was infallible proof of the mental superiority of men over women. Throughout his life he defended this hypothesis with the conviction of a zealot. Being the true scientist, he specified on his will that his own brain be added to his impressive collection. The postmortem examination elicited the interesting fact that his own brain weighed only 1245 grams."

AUGUST 1942

"I have no quarrel with—in fact, I commend—those individuals who want to store their cars for patriotic reasons. Nor will I argue about possible savings in gasoline and oil if cars are taken off the road. However, preservation is one thing and waste is another. Persons who are considering storing their cars for the duration to save tires should consider the fact that tires deteriorate when not in use. This is because normal operation of a car flexes the rubber and keeps it alive. Light— even artificial light—damages stored tires. All windows in the storage place should be blacked out and the car should be blocked up to remove the weight from the tires."

OCTOBER 1942

"'The problems we are going to meet after this war is over are two-fold,' said Mr. James F. Lincoln, President of the Lincoln Electric Company. 'First of all, there will be great unemployment because even if the call for goods is so great to keep all men employed, the reshuffling of these men to peace-time work will take some time. The second thing is that competition will have changed due to the existence of government financed organizations which will have much less over-head than those who have financed their own expansions. Of course, there are many unknown features. For instance, there is no way of knowing what kind of government we will have. It is rather obvious from history that it will be totalitarian. No bankrupt nation ever escaped totalitarianism. Perhaps all of our planning will go for naught because we will be told by bureaucrats what to do and how to do it.'"

NOVEMBER 1942

"Each week in *The Journal of the American Medical Association* there is a long series of obituary notices of deceased doctors. Each notice lists the dead doctor's life attainments. Naturally, some of these notices are long, some of medium length, others short, and it happens that, for reasons of appearance, the printer arranges the notices in order of length-longer ones preceding shorter ones. It occurred to a Brooklyn doctor that this weekly list, thus arranged, might provide an opportunity to determine 'what price success' in medicine. So he analyzed 30 such weekly lists and found that the average age of death of the first ten doctors on them was 64.6 years, while the latest ten doctors—they who had served faithfully but not gained prominence—lived on to 69.3 years. Thus the price of success proves to be about five years of a doctor's life."

FEBRUARY 1944

"The recent decline in the rate of discovery of new petroleum fields in this country has given rise to the question of what we can do to meet the demands of an air-minded and automotive post-war age. Great Britain, Germany, and Japan are making synthetic oil and gasoline. Now is the time to conduct a rigorous research program so that methods will be available to supply necessary liquid fuels from American coals when the petroleum supply begins to fall."

APRIL 1944

"Although the war has been responsible for many new inventions, it has added little to the world's store of fundamental knowledge, Dr. Frank B. Jewett, vice-president of the American Telephone and Telegraph Company, recently told members of the New York University Institute on Post-War Reconstruction. Progress in certain fields of scientific knowledge, he said, has been offset by a virtual cessation of research work in others that are not considered essential to the war effort."

FEBRUARY 1947

"Uranium metal could be used as an international monetary standard to replace the silver and gold that have traditionally set the world's standards of values. Atomic fission can convert a part at least of any mass of uranium directly into energy, and energy, the ability to do work, is suggested as a far more logical basis of economic value than any possessed by the precious metals. Uranium's hardness and the ease with which it oxidizes preclude its use in actual coins. However, the various proposals for international control of fissionable materials might lend themselves to an international paper currency backed by centrally controlled uranium metal."

JULY 1947

"Unfortunately for the development of the light car in the U.S., much of the public thinking has been concerned with 'keeping up with the Joneses.' General Motors and Ford have apparently shelved their plans for such cars, feeling it 'inopportune' to divert materials and man-power to the production of light cars which have high mileage per gallon of gasoline. Such moves leave Crosley Motors alone with the opportunity to develop a leading position in the low-priced car market."

DECEMBER 1948

"However wrong George Gallup, Elmo Roper and other pollsters may have been in their forecasts of the recent election (Harry S. Truman against Thomas E. Dewey), no social scientist believes that public opinion polling itself was thereby discredited as a useful tool. Science often learns more from mistakes than from successes. In this case, the polling fiasco of 1948 had at least two healthy results: 1) it demonstrated dramatically that polling is far from being an exact science (which apparently needed public demonstration) and 2) it will force more rigorous standards upon the polling business."

CHAPTER SEVEN

OCTOBER 1847

"The Association of American Geologists have just closed their annual meeting. Huge round rocks called bolders, found throughout different parts of our continent, have engaged a large share of their discussion, in accounting for their origin, where they have come from and by what means. It appears that the theory of their transportation is the 'Age of Drifts'—that this continent was once the bed of the sea and that these bolders were brought from the North Pole by icebergs. This theory has a drifty foundation."

JANUARY 1849

"Prout, in his Treatise on Disease, says about tobacco, 'Although confessedly one of the most virulent poisons in nature, yet such is the fascinating influence of this noxious weed, that mankind resorts to it in every mode that can be devised to insure its stupefying and pernicious agency. The severe and dyspeptic symptoms sometimes produced by inveterate snuff-takers are well known: and I have seen such cases terminate fatally with malignant diseases of the stomach and liver. Surely, if the dictates of reason were allowed to prevail, an article so injurious to the health and so offensive in its mode of employment would speedily be banished from common use."

MAY 1854

"Leverrier has recently read an interesting paper before the French Academy upon the asteroid planets, their eccentric orbits and irregularities. All of their orbits are especially characterized by eccentricities, and by considerable inclinations. It therefore follows that the hypothesis stated by Dr. Olbers—that the asteroids (some of which he discovered) were derived from the wreck of a larger planet that had exploded—is incompatible with the real truth, inasmuch as the forces necessary to launch the fragments of a given body in such different routes would be of such improbable intensity as to render it mathematically absurd."

APRIL 1860

"According to the *Annual of Scientific Discovery* for 1860, the researches made in the department of electricity, during the past year, have been most important; Messrs. Faraday and Grove, of England, occupying, as in years past, the most prominent positions as investigators. The results of the experiments instituted by the latter gentleman are exceedingly curious, and must be regarded as all but proving the truth of the modern theory which assumes that electricity is not, in any sense, a material substance, but only an affection (state) or motion of particles of ordinary matter. Thus he shows conclusively, by a great number of carefully instituted experiments, that electricity cannot be transmitting through a vacuum, and that in transmitting electricity through gaseous media, the facility of transmission is increased by a degree of attenuation in the media, but that when a certain point of attenuation is passed, transmission becomes difficult, and finally impossible."

APRIL 1861

"A lecture was recently delivered before the Royal Institution of Great Britain by Professor William Thomson of Glasgow on the subject of atmospheric electricity, in which he stated that by experiments with the air pump and 'vacuum tubes' for exhibiting the electric light, according as we obtain a vacuous space, it appears to be a conductor rather than an insulator. Professor Thomson said: 'We now look on space as full. We know that light is propagated like sound through pressure and motion. We know that there is no substance of caloric:

inscrutable minute motions cause the expansion that is marked by the thermometer; these stimulate our sensations of heat. Fire is not laid up in coal any more than in a Leyden jar, but there is potential fire in each. We can conceive that electricity is an essence of matter, but whatever it is, one thing is quite certain, electricity in motion *is heat.*'"

MAY 1862

"The theory of spontaneous generation was long since proposed to account for the origin of beings whose germs were too minute or too obscure to attract attention. One after another the different organisms supposed to arise from spontaneous generation have been proved to originate from germs. At present the question of spontaneous generation concerns only the origin of entozoa and those minute organisms that can be studied with the aid of the microscope, as molds (minute fungi) and Infusoria, both animal and vegetable. The common theory that the spores or germs of these minute organisms are constantly floating in the atmosphere ready to start into activity whenever they meet with a suitable nidus has found an able advocate in M. Pasteur, of the École Normale Supérieure in Paris, who has published in the *Comptes Rendus* a series of valuable papers on this subject."

APRIL 1863

"It is generally believed by men of science that the interior of the earth is a mass of molten matter, the heat of which is intense beyond that which can be produced by known artificial modes. Many of the rocks that form the crust of the earth appear to have been once in a fluid condition. Dr. Siljestrom, a Swedish astronomer, expresses it as his belief that the interior of the earth is occupied by currents of various degrees of heat, which mix with each other and attain a certain degree of temperature in the same manner as substances subjected to all the physical influences of the earth's exterior. In other words, the theory is that a mass of fluid, possessing different temperatures in different parts of its interior, must be subjected to a process of convection. The result is usually a change of volume in the entire mass of circulating fluid, causing eruptions like those of volcanoes."

MAY 1863

"Sir Charles Lyell, the distinguished geologist, infers from recent researches and discoveries of implements in various parts of Europe that man may have lived on the earth thousands of centuries before the era of his advent according to common belief. France, England, Denmark and Switzerland were once peopled by a race that used flint hatchets and arrow heads, like the old North American Indians. After them came a race that used implements of bronze; and again these were succeeded by a race that used implements of iron. In one case Lyell shows the section of an ancient hut that had been built on the Scottish seacoast. It had been submerged by the sea for so long a period that 60 feet of marine strata had formed over it, and after this, by some convulsion or gradual upheaval of the earth, it was elevated to its former position out of the sea. This hut affords evidence of having been erected in a far remote pre-historic period."

MAY 1864

"For some years past M. Pasteur, a distinguished French chemist, has been engaged in investigating the phenomena of fermentation and putrefaction and the results attained to by him constitute some of the most important contributions made to chemical science during the past few years. In the report of researches heretofore published. M. Pasteur claims to have proved that the effects hitherto attributed to the atmosphere of oxidizing and thus consuming dead organic matter are really dependent on the growth of infusorial animalculæ."

JULY 1864

"Harvey's theory of the circulation of the blood, or rather the causes of the circulation, is beginning to be disputed; for blushing, sudden paleness of the face, flushing and chillness of the body frequently occur without any disturbance or modification of the heart's action. The steady movement of the blood in the capillaries, the circulation through the liver without the intervention of any propulsive force, the fact that after death the arteries are usually found empty, among other things, cannot be accounted for on the hypothesis that the heart is the sole mover of the blood."

NOVEMBER 1865

"Professor Miller and Mr. Huggins have constructed an instrument with which they have compared the spectra of the moon and planets and some of the fixed stars, and even of the nebulae, with the spectra of the principal metals. They observe that 'the elements most widely diffused through the host of stars are some of them most closely connected with the constitution of the living organisms on our globe, including hydrogen, sodium, magnesium and iron. On the whole we believe that spectrum observations on the stars contribute something toward an experimental basis on which a conclusion, hitherto but a pure speculation, may rest, viz.: that at least the brighter stars are, like our sun, upholding and energizing centers of systems of worlds adapted to be the abode of living beings.'"

MARCH 1866

"All the facts of geology tend to indicate an antiquity of which we are beginning to form but a dim idea. Take, for instance, a single example—England's well-known chalk. This consists entirely of shells and fragments of shells deposited at the bottom of an ancient sea far away from any continent. Such a process as this must be very slow; probably we should be much above the mark if we were to assume a rate of deposition of 10 inches in a century. Now the chalk is more than 1,000 feet in thickness and would therefore have required more than 120,000 years for its formation. Moreover, we must remember that many of the strata now existing have been formed at the expense of older ones; thus all the flint gravels in the south-east of England have been produced by the destruction of chalk. This again is a very slow process. It has been estimated that a cliff 500 feet high will be worn away at the rate of an inch in a century. The Wealden Valley is 22 miles in breadth, and on these data it has been calculated that the denudation of the Weald must have required more than 150,000,000 of years."

OCTOBER 1868

"The theory of the origin of species by natural selection, which was first enunciated by Darwin 10 years ago and which has been so widely discussed, has undoubtedly been gaining ground among the most celebrated naturalists. When this theory was first propounded, it met both vehement opposition and ridicule. It was attacked by philosophers and wits, and formed the subject of many a lampoon and satire. It was denounced as opposed to the teaching of revelation, as a system of guesses which were not sustained by either facts or logic. But there was a vitality in the theory, and the conclusions of a man who fortifies his opinions with such a host of facts as Mr. Darwin brought to sustain his are not easily put aside. One after another the thinkers of the entire world have slowly been accepting the theory, until it may be doubted that any hypothesis is more nearly established upon a permanent basis."

FEBRUARY 1870

"The great German chemist Liebig has finally broken the silence with which he has borne the attacks upon his theory of fermentation on the part of many chemists during the past 10 years, and has come out with one of those exhaustive and convincing replies that recall the best days of his great intellect. The reticence he has observed has emboldened some of the younger chemists to disclose weak points in their attacks, the whole power and force of his argument is leveled at the French Academician and renowned champion of the new school, Professor Pasteur of Paris. It was Pasteur who announced nine years ago as a result of his experiments that Liebig's explanation of the action of yeast upon sugar was entirely without scientific foundation. Since that time Pasteur has had it his own way, and the views published by him have been fast gaining in popularity until they appeared destined to be accepted by a majority of scientific men everywhere. Liebig's paper is therefore a perfect bombshell in the camp, and as soon as the smoke has cleared up and the fragments have been collected, we shall probably have about as nice a fight as has been witnessed among chemists for many a day. We shall not fail to inform our readers of the progress of the controversy, if anything practical grows out of it."

AUGUST 1872

"Though we are compelled to think of space as being unbounded, there is no mental necessity that compels us to think of it as being either filled or empty; whether it is filled or empty must be decided by experiment and observation. Are the vast regions that surround the stars, and across which their light is propagated, absolutely empty? The reply of modern science is negative. The notion of the luminiferous ether must not be considered as a vague or fanciful conception on the part of scientific men. Of its reality most of them are as convinced as they are of the existence of the sun and the moon. The ether has definite mechanical properties. It is almost infinitely more attenuated than any known gas, but its properties are those of a solid rather than of a gas. It resembles jelly rather than air. It is the vehicle for star light, and without it stars could not be seen. If the ether has a boundary, masses of ponderable matter might be conceived to exist beyond it, but they could emit no light. Moreover, a body once heated there would continue forever heated. The loss of heat is simply the abstraction of molecular motion by the ether. Where this medium is absent no cooling could occur."

AUGUST 1875

"Immense boulders, called 'lost rock,' which have no resemblance to any mass of rock in their vicinity, are scattered over the northern part of the United States. Similarly, there are heaps of sand, gravel and cobble stones which form many of our ridges, knolls and hills and which are totally unlike any fixed rock near them. To explain the transportation of these wanderers from their homes various theories have been advanced, such as the effects of floods, or of powerful mud currents or of gas explosions that hurled rocks in all directions. These and many others, however, fail to satisfy the observed conditions. To Louis Agassiz belongs the credit of first attributing all these phenomena to glacial action. Agassiz conceived of a sheet of ice and snow extended enough to cover a continent. Having noticed that markings below glaciers in the Alps were the same as those found beneath the ice mass, he compared these with similar appearances in northern Europe and Asia, and made the bold generalization that all were due to the very same cause, and that one vast sheet of ice must have covered all the northern regions of the globe."

AUGUST 1876

"Is the universe composed entirely of hydrogen? There are many eminent chemists, Professor Cooke among them, who believe that instead of there being 64 elements there is but one. What force we shall employ to dissociate the elements and convert them into that primitive form we are at a loss to say as yet, but the spectroscope leads us to think that heat, if sufficiently intense, may accomplish it."

JULY 1877

"That the earth was at one time incapable of sustaining life, and that at some time in the course of events life began to be, no one doubts for a moment. It is also pretty generally admitted among scientific men that the beginning of life was in all probability a natural event; and that the earlier forms of life did not embrace the more complex types now existing, but were of simpler structure, perhaps not unlike the lowly organisms now studied under the microscope. Here the question arises: Was the beginning of life a phenomenon single and unique, and are the bacteria of today the unaltered descendants of the earliest forms of life? Or may life have begun, and may it still begin, at any time by the concurrence of suitable conditions?"

MAY 1881

"The distinction between wanton cruelty and the infliction of pain for humanity's sake is an important one, and the Society for the Prevention of Cruelty to Animals, more recently styled the Humane Society, certainly does not make it. The tender mercies of the foolish are often cruel, and of such a nature are those displayed by the officers of the society referred to when they seek by exaggeration and misrepresentation to stop all use of living animals for the scientific advancement of physiology and medicine. It shows a pitiful state of popular intelligence, feeling and judgment when legislatures can be persuaded by fanatics to pass laws making it a crime to pursue a line of scientific investigation that has been more fruitful than any other in knowledge helpful for the prevention and cure of disease. Pasteur's recent brilliant and most promising discoveries in connection with chicken cholera were no doubt made at the cost of considerable discomfort to a small number of guinea pigs, rabbits and barn-yard fowls, but the certain issue of those discoveries must be to prevent an incalculable amount of distressing and fatal disease among these animals, with a possible issue of infinite value to humanity in furnishing a clue to a right understanding and treatment of many of man's disease."

JUNE 1881

"We cannot tell whether electricity is some peculiar kind of substance or some modification or motion of ordinary matter. No phenomena have thus far been discovered that absolutely negate the notion that electricity may be a subtle, imponderable fluid. Another view, however, seems to carry a greater weight of opinion in its favor—that, namely, of Maxwell. He regards an electric charge as the establishment of a peculiar state of strain among the atoms of the charged bodies and in the medium between them. A discharge consists in the sudden relief of this strain by a giving way of the intervening medium, without necessarily implying any transfer of substance through it. In its application the theory is mathematically difficult, but it opens the way for the establishment of relations between electricity and the other physical agents, especially light and heat. There is certainly great probability that some hypothesis will yet be found that will include in one general theory all the physical agents—light, heat, gravity and chemical affinity, as well as electricity and magnetism. But the hour and the man have not yet come."

JUNE 1882

"Few objects in the heavens have been treated with such unmerited neglect as the Great Nebula in Andromeda. Notwithstanding its enormous extent, as followed by the Harvard 15-inch achromatic, of $2^1/4$

degrees in length and one degree in breadth, and its conspicuous brightness, which makes it readily perceptible to the naked eye, it has resisted all inquiry. The telescope has failed. The spectroscope gives some kind of reply, but it is an indecisive one. It precludes at once the idea of a simple gaseous condition such as that of the Ring nebula, or the Dumb-bell or the wonder in Orion. What is that at which we gaze, overspreading field after field of the telescope with soft yet vivid light? If it is gaseous, the gas is unknown or in some hitherto unknown condition. If it is stellar, how are its components so concealed that they cannot be isolated? If stars are there, they must be numbered by the hundreds of thousands, yet they must be of a lesser magnitude than we associate with the idea of a star."

APRIL 1883

"Is there more than one force? We speak of the forces of nature and classify them as heat force, light force, electric force, etc. Should we not speak of the force of nature as exhibited in heat, in light, in electricity, etc.? You ask, if there is but one force, why does it not always manifest itself in the same way? Simply because the conditions are different at different times and in different places. The convertibility of these expressions of force, the one into the other, is perhaps the strongest proof of their identity of origin. It is impossible to study closely the properties of matter and the various phenomena exhibited in its relation to energy without coming to the conclusion that all forms of energy, however they may appear to our senses, are the offspring of one parent."

MARCH 1884

"Prof. Charles A. Young of Princeton in a recent lecture asks: 'Is there any connection between the maximums of sun spots and terrestrial disturbances of any kind?' The average of the occurrence of sun spots is, from maximum to maximum, either 13.11 years or 9.11 years. It was said that in 1870 there was a maximum of sun spots and an extraordinarily hot summer, but it was forgotten that observations to be conclusive must cover the whole of the earth's surface and not a part only. It so happened that in 1870 at our antipodes in New Zealand there was at the time of our hot summer an intensely cold winter, something phenomenal for that region. But an unmistakable connection has been discovered between the sun-spot maximums and aberrations of the magnetic needle. It is well established that the period of a maximum of spots is coincident with a period of the greatest declination of the needle."

JUNE 1897

"Owing to the scarcity of right whales in northern waters, Newfoundland is about to follow the example of Norway in making humpbacks and fin whales, which are said to be found in immense numbers round the coast, the objects of systematic pursuit. The superintendent of fisheries has organized a fleet of small steamers, with harpoons and explosive lances, such as are used in Norway, to carry on the fishery. If the whalers of Newfoundland take many specimens, it might be worthwhile to try preparing its flesh for the market. If the prejudice against its use could be overcome, there is no reason why 'whale steak,' preserved and put up in tins, should not find ready sale."

JANUARY 1899

"Dr. G. B. Grassi for a long time had doubts on the connection between mosquitoes and malaria, owing to the absence of malaria from certain districts where mosquitoes abound. A careful classification of the various species of gnat has now led him to the conclusion that the distribution of certain kinds coincides very closely with the distribution of the disease. The common *Culex pipiens* is to be regarded as perfectly innocuous. On the other hand, a large species *(Anopheles claviger, Fabr.)* known in Italy as 'zanzarone,' or 'moschino,' is constantly found associated with malaria, and is most abundant where the disease is most prevalent."

MARCH 1903

"The great question of astronomy is the complete and rigorous test of the Newtonian law of gravitation. This law has represented observations so well during a century and a half that it is a general belief that the law will prove true for all time and that it will be found to govern the motions of the stars as well as those of our solar system. We know that the law of gravitation is modified in the motions of the matter that forms the tails of comets. There is an anomaly in the theory of Mercury which the law does not explain, and the motion of our moon is not yet represented by theory."

OCTOBER 1903

"Lord Kelvin made an interesting suggestion in connection with the perpetual emission of heat by radioactive substances. He said that if the emission of heat went on for 10,000 hours, there would be as much heat as would raise the temperature of 900,000 grammes of water 1 degree Centigrade. It seemed utterly impossible to Lord Kelvin that this would come from the store of energy lost out of a gramme of radium in 10,000 hours. It seemed, therefore, absolutely certain that the energy must somehow be supplied from without. He suggested that ethereal waves might in some way supply energy to radium while it was emitting heat to matter around it."

DECEMBER 1904

"A Dutch naturalist, Hugo de Vries, has just given the finishing stroke to the theory of natural selection, and has proposed in place of it another hypothesis which he calls "the theory of mutation." His main idea is the abrupt mutation of living forms, of which he has sought direct proof. The best one would be to find a plant that was actually in its period of mutation and that might beget a number of daughter plants in which there should shortly appear the characters of a new species. There would be more chance of finding a wild species undergoing a crisis of mutation among the species that present a great many subspecies; de Vries, therefore, experimented with 100 plants that satisfied this condition. He chose seeds from those which were distinguished by some peculiarity or deviation, like fissuration of the leaves, ramification of the spines, etc. Only one attempt fully succeeded, that which related to the Onagra, *Oenothera lamarckiana*. De

Vries cultivated it in his experimental beds from 1886 to 1900. In 1887 a new type made its appearance. In 1900, after eight generations, he had obtained, from 50,000 plants produced from his several sowings, 800 new individuals belonging to seven undescribed species. The new species appeared suddenly, without preliminary or intermediate forms. The result of these experiments furnishes a new and powerful argument in favor of the theory of mutation."

AUGUST 1905

"The question as to whether or not the earth carries the ether near it in its journey through space is one of very great theoretical importance, and the last word upon it has not yet been said. The results of Michelson and Morley's experiments with their interferometer can be accounted for most simply by supposing that both the earth and the ether near it are moving with the same speed—that is to say, that the earth drags the surrounding ether with it in much the same way as that by which a layer of air is carried by a projectile. Morley has recently varied his previous investigation with the object of testing whether the compensation which cancels the effect due to relative motion is complete in every case. It is the shrinkage of the base-plate of his apparatus which may come into play; and, besides improving the apparatus by increasing its sensitivity, he has changed the material of this plate from iron to wood. There is still absence of any indication of relative motion of earth and ether, and the proportional shortening must therefore be the same as in the previous experiments. It may at first sight seem unlikely that two such different materials should be equally affected. This must be determined by something more fine-grained than molecules, and this something must be essentially identical in both these bodies."

FEBRUARY 1907

"In 1903, Prof. Svante Arrhenius stated that the theory called panspermy, according to which the germs of organic life are conveyed through interstellar space from one heavenly body to another, had advanced greatly in probability from the establishment of the pressure exerted by light and other cosmic radiations. The theory was suggested after the repeated failures of many eminent biologists to discover a single case of spontaneous generation. A great difficulty of the theory had consisted in the apparent impossibility of conveying germs even from one planet to another in a time through which their life could be preserved. Most germs can be kept alive only a few years, although certain spores live for decades. By introducing the pressure of radiation as a motive power, however, these journeys can be reduced to 20 days' travel to the nearest planet, Mars, and 9,000 years to the nearest star, Alpha Centauri. But even these intervals may appear to be of formidable length, especially in view of the absolute dryness and intense cold and light to which the germs would be subject in transit. Arrhenius argues, however, that recent experiments by Roux prove that germs are not killed by light but by air. Furthermore, it has been demonstrated that intense cold is not injurious to all germs and that certain algae are not killed by being kept in a desiccator for over 20 weeks. With these objections removed, concludes Arrhenius, it would appear that interstellar space can be traversed, at enormous speed, by living germs which bring organic life to planets as soon as a crust capable of sustaining life has been formed."

JANUARY 1913

"It would appear that radium has landed geologists and biologists in a difficulty greater than that from which it was hoped it would deliver them. There is radium in the earth, and radium in disintegrating gives out heat. Therefore a once molten globe will cool down more slowly than if it contained no such independent source of heat. Lord Kelvin's calculations were made on the supposition that there was no source of heat except what the earth possessed as a molten globe. Hence we are at liberty to extend the time that has elapsed since the earth became the possible theater of geological change to 500,000,000, 1,000,000,000, or even more years ago. Radium has given us a blank cheque on the bank of time. But when the actual calculations were made as to how much radium known to exist in the outer shell of the earth would effect its cooling, this was found to be too great. It would, in fact, *raise* the temperature of the earth a fraction of a degree annually."

AUGUST 1913

"Air currents at a height of 50 miles above the earth are discussed, by J. Edmund Clark in the *Quarterly Journal of the Royal Meteorological Society,* on the basis of observations made at many places in southern England and northern France of the drift of a particularly bright and persistent meteor train seen on the night of February 22, 1909. Mr. Clark himself saw the train for 104 minutes. The most remarkable conclusions drawn by the writer relate to the velocity of the upper winds at various levels, as indicated by the movement of the tram. Thus it appears that between 49½ and 51 miles' altitude the streak lay in a west wind of more than 170 miles an hour, whereas at 51½ miles the current was almost from the east with a velocity approaching 200 miles an hour. These conclusions hardly agree with the prevailing conception of the stratosphere as a region of gentle winds."

SEPTEMBER 1913

"In a paper presented before the Birmingham meeting of the British Association for the Advancement of Science, Mr. C. R. Enock maintained that the economic problems before the world at the present time call for the establishment and exercise of a comprehensive and constructive science whose aim would be to evolve and teach the principles under which economic equilibrium in the life of communities may be attained. It was argued that the real science of living on the earth, or 'human geography,' the adaptation of natural resources and national potentialities to the life of the community, has never been formulated. The congestion of the population in towns, the desertion of the countryside, the high cost of living, low wages, unemployment and so forth are related phenomena, intimately connected with the conservation and development of natural resources. The axiom was advanced that the world is capable of supporting all its inhabitants in sufficiency, and its failure to do so is due to the non-emergence so far of an organizing science whose deliberations would be aloof from egoistic or partisan influences. It was affirmed that the teaching and operation of such a science are necessary if social security is to be maintained and civilization advanced; and it was suggested that to give effect thereto an institution should be established which would bear the same relation to the science of living as their corresponding institutions do to physical, geographical, medical and other sciences."

OCTOBER 1913

"The London *Times* correspondent in Munich reports that all Germany is obsessed with the idea of procuring meso-thorium for use as a panacea for cancer. This radioactive substance is obtained from the thorium waste in the manufacture of gas-mantles. For therapeutic use a tiny particle is inclosed in a silver covering pierced with minute holes: the box is placed upon the part affected with cancerous growth and is said to slowly but surely eradicate the disease, although leading physicians are disposed to reserve judgment on the subject."

SEPTEMBER 1917

"The many speculations that have been published concerning the origin of life on the earth and on any other bodies in the universe where it may possibly exist usually assume that, in some way or other, 'life germs' are transported across the gulfs of space from one planet to another. Thus it has been suggested that life may have been brought to the earth in meteors. One of the most recent suggestions is that minute 'life germs' may escape from the atmosphere of a planet in which life exists, just as molecules of the atmospheric gases are believed to escape from our terrestrial atmosphere, and may be driven by light-pressure to some world where physical conditions have become suitable to support life. While there is nothing essentially absurd in these hypotheses, it is not clear why their authors should take it for granted that life cannot originate *de novo* on a cooling planet."

OCTOBER 1919

"A recent paper by Mr. Harlow Shapley brings to a focus the long-standing disagreement between physical astronomers and geologists concerning the duration of solar radiation and the consequent age of the earth, and also the question of the age of the stars in general. The short time scale of the astronomers is based on the assumption that the sun's heat, flowing uniformly at the observed rate in all directions, comes from such recognized sources as gravitational contraction, the fall of meteorites, radioactivity, etc. The supply of heat from these sources could last only a few million years. The data of geology, however, particularly the recent work on sedimentation and on the radioactivity of rocks, are decidedly opposed to a short time scale for the sun and earth. Mr. Shapley cites a large amount of recent astrophysical evidence in behalf of the belief that the ages of the stars are probably several hundred times as great as was assumed in the older physical astronomy. Obviously existing hypotheses concerning the source of the energy radiated from the sun and stars need revision to bring them into agreement with the evidence in favor of an exceedingly prolonged life for these bodies."

JUNE 1920

"Perhaps the very experiment which started Einstein at work on his now famous 'theory of relativity' will be the means of demonstrating to the scientific world the correctness or incorrectness of his conclusions. The classical experiment performed by Michelson and Morley in 1887 and later repeated with better apparatus by Morley and Miller to determine 'ethereal drift' upon the assumption that the ether of space is a stationary sea through which our earth moves, is the experiment referred to. New interest is aroused by the statement made recently by Prof. Davton C. Miller, one of the original experimenters, that the work has never been carried far enough to be positive in its results. He has proposed moving the apparatus, which is still intact, to the top of Mount Wilson, near the famous Lick Observatory, and repeating the experiment there with the cooperation of the Lick scientists. Such an event would undoubtedly attract the attention of the entire scientific world, for although many scientists now recognize the probability that Einstein's theory will displace Newton's law of gravitation and indeed revolutionize our whole fundamental conception of the structure of matter and space, there are still many doubters."

OCTOBER 1933

"It is being asserted widely that science has at last abandoned its early materialistic taint, swinging strongly toward religion and mysticism. The furor started small in a corner of physical science. It was discovered a few years ago that things as small as electrons do not individually obey the law of cause and effect. This phenomenon is known as the principle of indeterminacy, or the principle of uncertainty, and its discovery came about when the German physicist Werner Heisenberg showed that it is impossible to ascertain both the position and speed of an electron; we can ascertain the one or the other singly, but not both. Evidently, then, the behavior of an electron is indeterminable. Right here a number of thinkers made false deductions and, as A. S. Eddington put it, 'science went off the gold standard.' What these thinkers failed to grasp was that mere indeterminability does not in itself establish indeterminacy. A thing may be indeterminable but not indeterminate. Nature knows what she is doing, and does it, even when we cannot find out. It did not take the mystics long to discover the principle of uncertainty. If we could no longer predict, at least in theory, the entire future of the universe, given the position and velocity of every particle in it, then perhaps there was freedom in it after all. The return of science to some sort of modern mysticism would be essentially a slip in man's hard-won progress away from one of his most ancient bad habits—that of ascribing to the supernatural whatever he did not yet understand."

JULY 1923

"If we are to believe the very logical arguments of the World Metric Standardization Council, there exists no good reason at all why the meter-liter-gram system has not been adopted by the people of the United States and Great Britain, except that deep-seated quality of human nature which causes us all to put our backs up and resist changes until they are forced on us. The greatest physical obstacle to overcome is the transition from the system of machine tools. Many American manufacturers, however, already are using the metric system of measurements today for the production of export articles.

What remains to be done is not so much to convince the average man of the desirability of the change on theoretical grounds as to demonstrate to him that he should contribute his share to making the change."

MAY 1925

"William Jennings Bryan, the recognized leader of anti-evolutionary thought, is ignorant of the facts of evolution and has no legitimate claim to popular leadership in such an issue. This was the statement made by Professor Edward L. Rice before the American Association for the Advancement of Science. Although Mr. Bryan recently joined this great association himself, and it is hoped that he will now become letter-perfect in the scientist's point of view, an analysis of his writings against evolution shows that he has not taken the pains to inform himself on the subject. Professor Rice suggests that Mr. Bryan should get down to facts rather than opinions and concludes that Mr. Bryan advances no new evidence but ignores or denies the data collected by scientists."

JANUARY 1926

"Professor Alfred L. Wegener, of the department of mineralogy and climatology at the University of Graz, says that millions of years ago the two Americas, as well as Europe, Asia, Australia, Antarctica and all the islands of the present-day world, were one continent centered around Africa. Tidal forces—the attraction of the sun and moon for the earth's solid mass (not ocean tides)—broke this super-continent up, and the pieces slowly dispersed in various directions, like the blocks of a great, flat cake of floating ice that is broken up by the waves. Some of these pieces, the present continents and islands, are drifting still, gradually moving away from the nuclear Africa. This theory is startling. To many it seems absurd. It may prove to be erroneous. It may gain final acceptance among geologists. At the present time it is strongly heterodox, but there is something about it that seems to captivate the interest of scientists."

APRIL 1928

"The average amount of uranium in the granitic rocks that form the outer layer of the earth's crust can be found with considerable accuracy. It appears to be about seven parts in a million by weight. Its decay must have produced a quantity of lead that must be somewhere in the same rocks. The total content of lead in these rocks, according to the best analyses, averages about 22 parts in a million, but the lead has not the right atomic weight. That it may be a mixture of various 'isotopes' having different atomic weights but the same chemical properties has long been suspected but was first proved by F. W. Aston only a few months ago. With the ingenious 'mass spectrograph,' in which a thin beam of electrically charged atoms is subjected to electric and magnetic fields so proportioned that atoms of each separate atomic weight are directed to a separate point on a photographic plate he has at last, after many vain attempts, attained many definite results with lead which show that there are three kinds of lead atoms with weights 206, 207 and 208. About 30 percent of all the atoms are of the first kind, 20 percent of the second and 50 of the third. Only lead of the first kind can be the product of the decay of uranium. The present amount of this is a little less by weight than that of the uranium itself. A simple calculation shows that it would have been produced by the decay of the uranium in 4,800,000,000 years. If the earth's crust had existed for longer than that, there ought to be more '206-lead' contained in the rocks in proportion to the uranium. We can therefore say, on the basis of this and other evidence, that the earth is more than one and a half billion and less than four billion years old."

FEBRUARY 1929

"Where is the solar system located within our galaxy. The first real idea of how far away from the center we are came when Harlow Shapley applied a new method of determining stellar distances: the photometric method. The story has often been rehearsed of how the true brightness of the Cepheid variables was found to depend on their period of pulsation. In addition the globular star clusters form a gigantic assemblage with its center about 50,000 light-years from the sun in the direction of the constellation Sagittarius. One region in particular, about eight degrees square, on the borders of Scorpio and Ophiuchus, has been found to contain no fewer than 450 variable stars. Twenty per cent of them lie at distances ranging from 21,000 to 30,000 light-years. The rest are fainter and range from 35,000 to 55,000 light-years, with a sharp concentration at about the middle of this interval. These stars lie within a few degrees of the direction of the system of globular cluster and at substantially the same distance. It appears, then, that there is actually an enormous mass of stars concentrated near this point that forms the central nucleus of our galaxy."

JULY 1931

"The world is not as familiar with the name of Willem de Sitter, the Dutch astronomer-cosmologist, as it is with that of Albert Einstein. Readers of science often encounter the term 'the Einstein universe,' 'the de Sitter universe' and so on. Such expressions refer to concepts of the size, shape and general nature, also the finiteness or infiniteness, of the totality of things. Of course, no existing telescope can penetrate the entire extent of the universe, and therefore these concepts are based partly on inference. In de Sitter's universe, which is finite, space is curved or bent, not so much because of the presence of matter, as in Einstein's universe, but inherently. It is an unstable universe, expanding or contracting. Recent research by E. P. Hubble of the Mount Wilson Observatory, indicating an expansion, favors de Sitter's concept and has caused Einstein to revise his own."

JANUARY 1932

"The assumption that it would be quite as easy to breed human beings up to the level of our 'best' people as it is for men to breed cattle or horses of desirable qualities is not only unscientific; it is ridiculous. No matter how normal and moral and noble we happen to be individually, we are carrying within us a heritage of weak genes that are masked by genes of better quality. Dr. Sewall Wright, the University of Chicago zoologist, has bred guinea pigs all kinds of ways to uncover their weak genes. He did this by making brother-and-sister matings and by building up 'pure' lines. In such lines all kinds of potential or inherent weaknesses develop, and the lines, so long as the brother-and-sister crosses are continued, breed true for their weaknesses. It might be possible, given many years, to breed a colony of guinea pigs in such a manner as to secure nothing but pure-line progeny. That done, the animals with weaknesses could be killed or allowed to die out, and the valuable ones could be mated to secure a finer race of guinea pigs than ever existed since the world began. It would be all but impossible to apply such a method to human beings and to breed monstrosities deliberately, even if we knew the factors governing intellectual inheritance (which we do not), or had agreed on the type of human being that would be the best and most valuable (which we probably cannot)."

JANUARY 1933

"The most recent atomic research offers the heartening possibility of ultimately amalgamating the wave and particle concepts of matter. As a matter of fact, there are *no* experiments which prove that matter possesses *all* the properties of *either* a wave *or* a particle. Accordingly Werner Heisenberg has suggested that the two mental pictures that experiment conjures out of our imagination—one of particles, the other of waves—are both to be viewed simply as incomplete analogies arising from our temporary inability to describe matter in everyday language. Although we cannot draw a satisfactory picture of the atom as it is conceived by wave mechanics, the mathematics of the theory enables us to do all the things that were possible with the Bohr model of the atom. In addition it appears that what were heretofore contradictions are removed, fundamental points are refined and made more precise, and the number of assumptions necessary to attain those ends are reduced to a minimum. From this point of view wave mechanics signifies not so much a radical change as it does a welcome and highly significant evolution of the existent atomic theory."

MARCH 1936

"There is a great tendency today to add vitamins to foods. The wisdom of this is doubtful, for the ordinary well-balanced diet supplies all the vitamins that human beings usually need. The American Medical Association not long ago denounced the crude and unscientific character of vitamin therapy. It said there was no more reason for people to take varied dosages of several vitamin concentrates incorporated in food or drug products than for them to dose up on any other individual, unrelated dietetic components."

APRIL 1943

"The question of whether ill health can result from lead piping for household water supply has no categorical answer. The following is the reply given to a physician by *The Journal of the American Medical Association*. 'The amount of lead absorbed by most waters is negligible. Lead piping is effective in forming an insoluble coating of salts which inhibits its solution. It is *only when the water supply* is acids, particularly because of organic acids, that it is a potential danger. It may also dissolve when different metals are used in the plumbing, when galvanization may play a part. Water with highly solvent properties will dissolve some lead from a pipe on standing. The length of standing and the temperature of the water will influence the final concentration, but the actual quantities of lead will be small.'"

JULY 1943

"If, as appears to be probable, vegetation exists on Mars, life has developed on two out of the three planets in our system where it has any chance to do so. With this as a guide, it appears now to be probable that the whole number of inhabited worlds within the Galaxy is considerable. To think of thousands, or even more, now appears far more reasonable than to suppose that our planet alone is the abode of life and reason. What the forms of life might be on these many worlds is a question before which even the most speculative mind may quail. Imagination, in the absence of more knowledge of the nature of life than we now possess, is unequal to the task. There is no reason, however, against supposing that, under favorable conditions, organisms may have evolved which equal or surpass man in reason and knowledge of Nature—and, let us hope, in harmony among themselves!"

CHAPTER EIGHT

AUGUST 1846

"It is well known that there is a constant emission of hydrogen from the decomposition of various substances; and that this gas, being buoyant, has a tendency to rise to the surface of the atmosphere. According to one view, there is therefore no doubt that immense quantities of this inflammable substance abound in the upper regions, and that a spark of electric fire would envelope the world in flames. The only circumstance preventing such conflagration is that the region of excitable electricity is several miles below that of the inflammable air."

JUNE 1847

"A gentleman in Glasgow, Scotland, suggests a ready method to prevent sailing vessels from being consumed by fire. Every vessel should carry as ballast a quantity of chalk. In the event of fire in the hold, by pouring diluted sulphuric acid onto the chalk, such a quantity of carbonic acid gas (carbon dioxide) would be generated as would effectually put out the flames."

APRIL 1848

"A communication to the Paris Academy of Sciences from Monsieur Pallas suggests that the greater number of nervous affections are occasioned by the excessive influence of atmospheric or terrestrial electricity. He states that by adding glass feet to bedsteads and isolating them about eighteen inches from the wall, he has cured the patients sleeping upon them of a host of nervous affections."

DECEMBER 1848

"The huge dam over the Connecticut River at Hadley Falls, Mass., was completed on the 16th of last month, and the day of its completion was the day of its doom. A great number of people had assembled to see the dam filled, and the waters of the Connecticut arrested in their course. But from the first, imperfections were discovered in the work, and a breach, small at first, widened with great rapidity, until about three-fourths of the embankments burst away before the mighty mass of angry waters. The dam was constructed of immense timbers, fastened to the rocky bed of the river with iron bolts. Fault must be attributed to the principle of its construction."

MARCH 1849

"The *Presse*, of Vienna, Austria, has the following: 'Venice is to be bombarded by balloons, as the lagunes prevent the approaching of artillery. Five balloons, each twenty-three feet in diameter, are in construction at Treviso. In a favorable wind the balloons will be launched and directed as near to Venice as possible, and on their being brought to vertical positions over the town, they will be fired by electro magnetism by means of a long isolated copper wire with a large galvanic battery placed on the shore. The bomb falls perpendicularly, and explodes on reaching the ground.'"

APRIL 1852

"American Geographical Society read a very interesting paper from the Rev. Mr. David Livingstone, a missionary in South Africa. The people here are strongly developed, but peaceful. The Baloc tribes melt large quantities of iron, and are very good smiths. The people are all aware of the existence of a God, and seem to be informed in regard to future life, and rewards and punishments. The Portuguese slave traders begin to penetrate there. About two years ago some traders came into the Chobe region, but the people were not inclined to the business. The price of a boy was about eight or nine yards of calico or baize cloth. Mr. Livingstone proposes to send his family home and go himself as a missionary to reside in the heart of the country."

JULY 1854

"The eminent astronomer Herschel has suggested that the sun may be inhabited, and that between its luminous atmosphere and its surface there may be interposed a screen of clouds, whereby its inhabitants may no more suffer from intense heat than those who live in our tropical regions. This may be so, as we all know how much the heat of the sun's rays, in the hottest days of summer, are modified by an interposing cloud or 'a swift passing breeze.'"

SEPTEMBER 1855

"Captain William Allan, of the British Navy, has published a book advocating the conversion of the Arabian desert into an ocean. The author believes that the great valley extending from the southern depression of the Lebanon range to the head of the Gulf of Akaba, the eastern branch of the head of the Red Sea, has once been an ocean. It is in many places 1,300 feet below the level of the Mediterranean, and in it are situated the Dead Sea and the Sea of Tiberias. He believes that this ocean, being cut off from the Red Sea by the rise of the land at the southern extremity, and being fed only by small streams, gradually became dried by solar evaporation. He proposes to cut a canal of adequate size from the head of the Gulf of Akaba to the Dead Sea, and another from the Mediterranean, near Mount Carmel, across the plain Esdraelon, to the fissure in the mountain range of Lebanon. By this means, the Mediterranean would rush in, with a fall of 1,300 feet, fill up the valley, and substitute an ocean of 2,000 square miles in extent for a barren, useless desert; thus making the navigation to India as short as the overland route, spreading fertility over a now arid country."

OCTOBER 1861

"For several years a Pneumatic Dispatch Company has been in operation in London, pipes of a few inches in diameter being laid through which small parcels were sent to various parts of the city. The company, finding the system to work well, have decided to enlarge the tubes to a hight of two feet nine inches and to a width of two feet six inches, and ultimately extend their system throughout the whole metropolis. Trucks six or seven feet long are sent through these tubes with loads of one or two tons. But the most interesting incident is that *two gentlemen have already ridden through the tube on one of the trucks,* thus perhaps inaugurating a new system of passenger traffic."

MAY 1866

"On the 17th, the House of Representatives passed laws which legalize the use of the metrical, or decimal, system of weights and measures in the United States. The important movement met with no opposition, and it is probable that within a few days the action of the House will be confirmed by the Senate, when the metrical system will become the law of the land. In the beginning the use of the system is not to be compulsory but optional with the people. As soon, however, as it becomes well enough understood it will entirely supersede the present system."

JULY 1866

"Prof. Henry Draper writes in a new periodical called the *Galaxy*: 'The celebrated nebular hypothesis of Herschel and Laplace, which assumes that our solar system was at one time a gaseous mass, extending beyond the orbit of the farthest planet, Neptune, has been very freely discussed and has received much adverse criticism. Many strong objections have been urged against it, but the spectroscope confirms it. On applying the spectroscope to the investigation of the irresolvable nebulae. Huggins finds that some of them present the spectra characteristic of an ignited gas, that is, of a flame. The Fraunhofer lines in that case are bright instead of dark, as in the solar spectrum, and the evidence is of a very tangible and unmistakable kind. There are, then, in space masses of ignited gaseous matter of prodigious extent, shining by their own light, containing no star and resembling the nebulae, which the nebular hypothesis declares to have been the original state of the solar system.'"

MAY 1867

"At a crowded meeting of the members of the Royal Society of Scotland. Professor Sir William Thomson presented a communication based upon the admirable discovery by Helmholtz of the law of vortex motion in a perfect liquid. That is to say, in a fluid destitute of viscosity or friction Helmholtz has proved mathematically an absolute unchangeability in the motion of any portion of a perfect liquid in which the peculiar motion he calls wirbel bewegung motion has been created. Professor Thomson therefore boldly throws down the gauntlet by condemning 'the monstrous assumption of infinitely strong and infinitely rigid pieces of matter and suggests that Helmholtz rings are the only true atoms. A full mathematical investigation, said Professor Thomson of the mutual action between two vortex rings of any given magnitudes and velocities passing each other in any two lines so directed that they never come nearer to each other than a large multiple of the diameter of either, is a perfectly solvable mathematical problem and the novelty of the circumstances presents difficulties of an exciting character. Its solution will become the foundation of a proposed new kinetic theory of gases."

NOVEMBER 1868

"The velocipede mania is beginning to set in, and with the opening of the spring months we may expect to see our parks and highways thronged with this cheap and agreeable substitute for the horse. The two-wheeled velocipede is not exactly the thing wanted for general use, as it will be somewhat difficult for novices to keep upright upon it. A nicely adjusted vehicle with a double hind wheel would be most desirable for all classes."

MARCH 1869

"A company has been organized in London to tunnel from the Post Office to the marble arch entrance on Hyde Park. The trains of the proposed line are to be drawn by wire ropes from fixed engines at each end, so that the air of the tunnel will not be poisoned by the smoke and vapors of the locomotives, since there can be no collisions, trams will start every two minutes. In the opinion of competent engineers the substitution of locomotives for ropes was a mistake, whether regarded from the scientific or the economic point of view. Improved means of communication in cities is one of the greatest necessities of the day. We want, if possible, to get rid of the surface roads."

FEBRUARY 1872

It has been estimated that the domestic consumption of ice in New York, Brooklyn, and vicinity is a tunage equal to that of the domestic consumption of coal. Whether this estimate be too large or too small, it is certain that ice has become an article of almost as universal demand as coal. The comfort, economy, and convenience secured by the use of ice is so great that it may now be classed as one of the indispensable articles of city consumption. Its harvest and supply has grown into an enormous business, which at times assumes the attitude of a merciless monopoly, and, in the absence of effective competition, is enriching the companies that conduct the business. It has also called into existence the use of improved appliances for cutting and storing ice, by which advantage can be taken in even short seasons, to secure the quantity needed. It has been proposed to use steam appliances for cutting the ice, but at present the ice plow drawn by horses is the principal method employed.

The ice plow used for cutting the ice is not very unlike an ordinary plow. For the solitary pointed blade are substituted several long, sharp prongs or teeth, which act saw fashion, and are so adjusted that the ice is cut but half through. When thus cut, the blocks are easily separated from each other. The furrows are opened in parallel lines, giving a surface dimension to the blocks of two and a half feet by two feet. As the plow passes over a small area, the men, furnished with long poles terminating in strong iron hooks for the purpose, haul the blocks to the source of the canal, where, after twenty-five or thirty blocks are collected attachments are made, and another horse tows the blocks to shore through the channel previously cut. This channel often extends a mile from the shore, and forms a canal for the transport of the ice rafts.

While the harvest is going on, the ice field presents a busy scene, being dotted all over with laborers. The work is often, in short seasons, pursued by moonlight, and pushed unremittingly to secure the crop before a thaw destroys it.

When the ice reaches the bank, it is hoisted up inclined planes and down others, into the storehouses.

HARVESTING ICE ON THE HUDSON.

The ice houses are constructed with every regard for atmospheric changes, and are models of simplicity. The large one contains six rooms, four of which are 75 x 50 feet in area, and of an altitude sufficient to allow a packing of ice 30 feet high, and an open space of 20 feet for air. The two remaining rooms are 150 x 50 feet in dimension, and the entire building has a capacity of 48,000 tuns of ice. The walls of the houses are double, and filled in with sawdust and tan. At the end of the houses nearest the canal, the apparatus for raising the blocks is constructed, extending from the water to the roof. From a distance this looks like two heavy ladders laid upon an inclined plane, each furnished with a pair of hand rails. At the base of each are two pairs of wheels, over which pass endless chains, stretching to the summit. To these, bars are attached, at a respective distance of six feet, which with the chains from the "apron." On a level with each floor of the building, a platform connects the plane and door sill, on which the blocks are deposited in order to fill each story in succession.

As the ice reaches the base of the plane, the blocks are pushed one by one close to the lower pairs of wheels. as the chains force the bars along' they catch the blocks and carry them up to the second floor, A strong staging is built in the interior of the house, extending to the different rooms, on which the blocks are pushed to the apartment intended for their storage. From this staging an incline extends to the highest layer of ice. As the blocks are deposited on the platform, the first is pushed towards the incline in the nearest apartment, the second in the next, and so on, the entire floor filling up evenly.

In order to prevent the blocks crushing against each other, as they slide from the plane, a number of large headed nails called "scatchers" are driven in it, and greatly diminish their velocity and render their "shooting" of short rang.

Five per cent of all ice received into the building becomes useless by cracking and scratching. After a layer of blocks is completed, the workmen shovel from the surface all the loose pleces and the snow, and throw them out of the building through the high, narrow air passages.

MAY 1874

"Babbage, in speaking of his analytical engine, has suggested that a machine might be made that would play a game of combination, such as draughts, provided the maker of the machine himself would work out perfectly the sequences of the game. Professor Fairman Rogers finds that the sequences of tit-tac-toe are easily tabulated, and hence an automaton may be made that will play the game as follows: The opponent to the automaton makes the first move in the game and in so doing causes a certain cylinder to change its position. This causes the automaton to make that play which the proper sequence of the game requires. The next play of the opponent moves another cylinder, and so on throughout the sequence. If the player plays perfectly, the game will be drawn; if the opponent makes a mistake, the automaton will take advantage of it and win the game."

DECEMBER 1874

For many years the project of building a railway tunnel under the bed of the Hudson river, between New York and Jersey City, has been discussed, its importance and feasibility agreed upon, and its successful completion, upon paper, established. Only two things have been lacking for the actual realization of the work, namely, the money to build with, and the company of individuals enterprising and bold enough to assume the risks incident to such a task.

The bed of the Hudson, at New York, is a treacherous substratum, so far as tunneling is concerned, being porous, leaky and lacking in firmness. All engineering experience in the construction of works in such soils has shown that their prosecution is attended with unusual risk and cost. But now comes along a new and enterprising engineer from California, Mr. D. C. Haskin, inventor of a new Improvement in the Art of Tunneling, expressly designed to make difficult works of this kind easy, patented February 3, 1874. Mr. Haskin has organized a strong and wealthy company for the trial of his improvements, and the first essay is to be made upon the Hudson River tunnel, work upon which has recently been commenced. The vertical shaft has already reached a considerable depth. It is located near the river shore at the foot of 15th Street, Jersey City, and from thence the tunnel will extend across under the Hudson river to or near the foot of Canal Street in New York, thence up Canal Street to a connection with the Broadway Underground Railway.

The greatest depth of water on the Hudson River over the tunnel will be about 100 feet; the total width of the river, 4,000 feet. The actual length of the horizontal tunnel, however, will hardly be less than 6,000 feet.

We will now describe this New Art of Tunneling, premising, however, that the failure of the plan, which we consider inevitable, will not, necessarily, stop the construction of the tunnel, as the air-compressing apparatus, which is the principal item of expense, will be useful in whatever method may be hereafter adopted.

Within the tunnel, a short distance back of the heading where the laborers are at work, is an air lock, composed of an iron cylinder, having entrance valves at each end, so arranged that when one is opened the other closes, thus permitting egress or ingress to the front. Above the lock is an airtight packing, while below the cylinder is a packing or filling of earth. When the air lock is duly set and sealed within the tunnel, compressed air is driven to the heading in front of the air lock, through air pipe. The excavated earth will be discharged from the heading, by the air pressure, through the pipe, and delivered into boats or other suitable receptacles at the ground or river surface, in the manner commonly practiced in sinking caissons.

In carrying on the work, the laborers will excavate a chamber in the earth in advance of the finished masonry, which will then be carried forward, while the men dig out a new space in advance, and so on until the tunnel is completed. Any loose boulders, stones, earth, quick sands, or water, encountered in the roof or walls of the heading, are to be held up and prevented from caving-in upon the workmen by the air, like flies upon the ceiling. The clumsy, costly caissons, shields, and other appliances, heretofore deemed necessary by cautious engineers, are discarded in this New Art. It is, indeed, a new wrinkle in the science of engineering.

But we think the statement of the patent, that only 50 lbs. air pressure will be required, must be a mistake. Several cyphers have evidently been omitted from the figures, perhaps by a blunder at the Patent Office. A cubic foot of air weighs only 0.075 of a pound, while a cubic foot of stone weighs 165 lbs. To buoy up such a stone in air, requires a corresponding density of the air: which involves the compression of 2,200 cubic feet of air into every cubic foot of air contents within the heading, of a pressure of 33,000 lbs. to the square inch.

Our author makes another rather incongruous statement in his patent. He says: "In case a jet seam or small stream of water is encountered, I supply a temporary shield of canvas, leather, or other light flexible integument to the wall, against which the pressure instantly forces it and seals the leak."

Water weighs only $62^{1}/_{2}$ lbs per cubic foot, or less than half the weight of granite. If the direct air pressure, against the loose earth, sand, and stones, is sufficient to prevent their downfall in the excavation, surely no streams of water can come in, and the leather will be unnecessary.

FEBRUARY 1875

"From a quotation in the London *Medical Record,* we learn that M. Philbert states that the principal measures for reducing obesity come under four heads: 1. *Régime;* 2. Hygiene; 3. Exercise and Gymnastics; 4. Waters with sulphate of soda. The basis of the *régime* rests on the prevention of the introduction of carbon into the system, or on favoring its transformation, and augmenting the amount of oxygen. The food must, therefore, be non-nitrogenous, varied with a few vegetables containing no starch, and some raw fruit. But the temperament of the patient must be kept in view. The lymphatic should have a red diet, beef, mutton, venison, hare, pheasant, partridge, etc., and the sanguine should have a white diet, veal, fowl, pigeons, oysters, etc. Vegetables, not sweet or farinaceous, may be allowed: grapes, gooseberries, apples, etc. *Café noir,* tea with little sugar and the addition of a little cognac, may be used."

JANUARY 1878

"The proposed pneumatic despatch system in Berlin will comprise 26 kilometers of tubing and fifteen stations. The bore of the tubes will be 65 millimeters. They will be of wrought iron and will lie about a meter below the surface of the ground. The letters and cards which are to be forwarded have a prescribed size, and are enclosed in iron boxes or carriers each of which can hold twenty. From ten to fifteen carriers are packed and forwarded at a time, and behind the last is placed a box with a leather ruffle, in order to secure the best possible closure of the tube. The exhausting machines and apparatus required for the transmission are situated at four of the stations. Both compressed and rarified air, or a combination of the two, are employed in propelling the carriers. Steam engines of about 12 horse power are used in condensing or rarifying the air. Each of the four main stations has two engines, which drive a compressing and exhausting apparatus. Large reservoirs are used for the condensed and rarified air. The tension of the condensed air is about three atmospheres, and that of the rarified about 35 millimeters of mercury. The condensed air, heated to 45° C. by the compression, is cooled in the reservoirs which are surrounded with water. The velocity of the carriers averages 1,000 meters per minute, and a train is despatched every 15 minutes. Each of the two circuits, into which the system is divided, is traversed in 20 minutes, including stoppages. The entire cost of the enterprise is estimated to cost 1,250,000 marks."

APRIL 1881

The utility of a really practical calculating machine can scarcely be overestimated. A great deal of time has been devoted to this subject, and no little money has been spent in endeavors to perfect a usable machine of this character; but hitherto the machines have been too complicated, too bulky, and too, expensive.

A short time since Mr. Ramon Verea, of 88 Wall street, New York city, patented a calculating machine involving an entirely new principle. It is comparatively simple and inexpensive, and is very compact. This machine cannot be intelligibly explained without engravings, but it may be stated that the essential features of the invention are a series of prisms perforated with holes of different sizes, and a series of tapering prisms which enter the holes more or less according to the size of the hole.

With this machine Mr. Verea can not only add and subtract readily, but he is able to perform multiplication and division with equal facility.

DECEMBER 1882

"The English sparrow was first introduced to this country in 1862. The importation was made solely in view of the benefits to result from the birds' immense consumption of larvae. The only positive result of their introduction is that the measuring worm, which formerly infested all our vegetation, is now very nearly extinct through the instrumentality of the sparrow. The amount of havoc in our wheat fields created yearly by the sparrows, however, is enormous. Their forwardness and activity have driven all other birds from where they have settled, so that the hairy caterpillars, which sparrows do not eat and which used to be extensively consumed by other birds, are now greatly on the increase."

OCTOBER 1883

For many years the popular ideal has been that whenever the genius of man should overcome the barrier to commerce which nature has placed at the American Isthmus in Panama, it would have to be accomplished by a ship canal. However, in laying before the world a plan of a ship railway, Mr. Eads offers no speculative project, but the well considered design of a capable and experienced engineer, a project contemplating the hauling of great ships over land from one sea to another. The proposal consists of nothing more novel than two marine railways joined by a few miles of many-railed roadbed of easy grades.

In transferring a ship from the harbor to the upland track, the cradle or ship-car will have to be backed down to the harbor end of the basin under water by means of a stationary engine. The ship will then be floated in, so that her keel will rest over the cradle. When entirely out, two powerful locomotives 5 times as large and powerful as ordinary freight engines, will be hitched on to haul the massive load to the opposite sea.

Mr. Eads is confident: it is possible to build and equip a ship railway for one-half the cost of a canal, that such a ship railway can be built in one-third to one quarter of the time needed for a canal, that a greater number of vessels per day can be transported on the railway than would be possible through the canal, and that the cost of maintenance of the ship railway will be much less than that of the maintenance of a canal.

The great Ship Railway will doubtless be in actual operation before three years have expired.

JANUARY 1884

"We are pleased to find that increased attention is being paid to the question of the physical training of young and growing girls. The Swedish physical exercises have found general favor, while many games and athletic pursuits are now allowed which formerly were proscribed by prudish schoolmistresses and timid mamas. When the girl is naturally healthy, little is wanted but to encourage ordinary systematic exercise being taken daily. This should consist of certain gymnastic exercises, which are to be practiced each day as part of the school work, supplemented by such games as lawn tennis, rounders, golf, etc. Swimming is an exercise that every girl should indulge in, and it ought to be taught systematically at all our girls' schools. Until recently dress proved a great barrier to free exercise of the limbs and body, but the introduction of a more sensible costume for the playground will in the future, it is to be hoped, remove the disadvantage. The costume in use consists of a short skirt of blue serge, draped with a crimson scarf, blue jersey, short trousers and long stockings."

SEPTEMBER 1886

"At the Baldwin Locomotive Works there are in course of construction four locomotives, designed to be run by soda, which takes the place of fire under the boiler. Inside the boiler will be placed five tons of soda, which, upon being dampened by a jet of steam, produces an intense heat. When the soda is thoroughly saturated (in about six hours), the action ceases, and then it is necessary to restore the fuel to its original state by forcing through the boiler a stream of super-heated steam from a stationary boiler. That drives the moisture entirely from the soda, when it is again ready for use. The engines, which are to be run on the streets of Minneapolis, will readily draw four light cars."

JANUARY 1887

"How to dispose of the snow: this is one of the most serious problems connected with the comfort and convenience of a great city in this latitude. A new method proposed by S. D. Locke of Hoosick Falls, N. Y., would utilize the steam plants existing in most cities to melt the snow. Underneath the surface gutter he proposes to construct a sub-gutter that connects directly with the sewer and that is covered with a grate, underneath which steam pipes are carried in racks. The snow can as quickly be moved by horse scrapers and brooms into the gutters as the streets can now be swept."

OCTOBER 1888

"The way in which railroad officials keep track of their freight cars, which are run thousands of miles over other railroad lines, has no doubt excited the wonder of many. Nearly all the great roads employ a corps of what are known as 'lost car searchers' or 'tracers.' Every freight car is numbered and used for a certain purpose, and whether it be a 'gondola,' or flat open car, or a box car, it can be traced from one end of the country to the other. At last one great trunk-line road has dispensed with the searcher in favor of a large force of clerks, with the telegraph and telephone as auxiliaries."

JANUARY 1889

"Subscribers to whom are rented phonographs can have left at their door every morning the waxy tablets known as phonograms, which can be wrapped about a cylinder and used in the phonograph. On these tablets will be impressed from the clear voice of a good talker a condensation of the best news of the day, which subscribers can have talked back at them as they sit at their breakfast tables."

SEPTEMBER 1889

"The Boynton bicycle engine, suggesting a very radical change in railway construction, arrived in New York last week from Portland, Me., where it was built. It weighs twenty-two tons, and came on a truck attached to the rear of a regular train. The engine and train are to be kept on their single track by upper wooden guiding beams supported fifteen feet above the track by a bridge-like skeleton frame arching above the roadway. This form of construction is designed to reduce the weight of the cars, both passenger and freight, relative to the load carried, and to save power lost in rounding curves, it being intended to so balance the train that there will be but little strain on the top guiding rail."

APRIL 1890

"An attachment for typewriters, by means of which the shifting of the characters and the spacing may be effected without using the hands, has been patented by Messrs. Reuben Durrin and Rosecrans Sheldon. To throw the capital characters into printing position the operator presses a knee against one side lever, pressing the opposite side lever when it is desired to space, while to throw the figures into printing position the central lever is pressed by the knee, the latter lever being adjustable to any desired height."

JANUARY 1893

"The Refuse Disposal Company, London, has lately published a pamphlet on the question as to the practical means by which the dust refuse of towns can be utilized for electric lighting purposes. The company claims that 20,000 tons of house dust, if treated as they suggest, and burnt in suitable boilers, might be made to produce as much as 5,600,000 indicated horse power hours, equal to an engine of 1,183 indicated horse power working for 4,734 hours, for electric lighting."

OCTOBER 1893

"Efforts have been made to teach a child how to swim by supporting him in the water and causing him to effect the motions of natation. This is the most practical process. Its inconvenience is that it necessitates the presence of a teacher with each pupil, and, in a large class of children, the teacher cannot occupy himself with each of them for a very long time. Mr. Devot has been able to overcome all the difficulties of the preceding method in a very ingenious manner. His apparatus permits the pupil to learn in conditions entirely identical with those that present themselves when he tries to sustain himself alone in the water. The apparatus is in use among the pupils of the Michelet Lyceum, who have been the first to benefit from the invention of their master, Mr. Devot."

JUNE 1895

"The days of ballot box stuffing and other modes of cheating at elections appear to be numbered. Inventive genius has provided machinery that will not lie and will not allow deception at the polls. As soon as the voter has recovered from the shock of the sudden and rather awful imprisonment in a chamber of steel, he is able to realize what is expected of him. Inside the voting machine, names of the candidates of the democratic party are printed upon a yellow background, candidates of the republican ticket upon a red background, and prohibition candidates upon a blue background. To the right of each name is a little knob which he must press in order to register his vote—the machine does the rest."

OCTOBER 1895

"The telephone newspaper organized at Pesth, Hungary, has now been working successfully for two years. It is called the Telephone Hirnondo, or Herald, costs 2 cents, like a printed paper, and is valuable to persons who are unable or too lazy to use their eyes or who cannot read. A special wire 168 miles long runs along the windows of the houses of subscribers, and within the houses long, flexible wires make it possible to carry the receiver to the bed or any other part of the room. To fill up the time when no news is coming in, the subscribers are entertained with vocal and instrumental concerts."

NOVEMBER 1896

"An immense crowd assembled near the Hotel Metropole, London, November 14, to witness the departure of the motor carriages for their race to Brighton, 47 miles. The occasion of the race was the going into effect of the new law which opens the highways to the use of the motor carriages and doing away with the antiquated laws which have hitherto obtained. It is a curious fact that under the old law self-propelled vehicles were not allowed to go faster than six miles an hour and had to be preceded by a horseman waving a red flag. Nearly fifty carriages started in the race; it is a great satisfaction to know that the race was won by the American Duryea motor wagon. The distance was covered in four hours."

JULY 1902

"Now that the submarine boat has fully justified its existence as a potential fighting factor which will exert a far-reaching influence upon naval battles of the future, efficacious destroyers are being sought for the purpose of nullifying its power and operations. A destructive machine for this purpose has been contrived by an English inventor, Mr. Gardner of London. The basis of his apparatus is an application of the transmission of ether waves. Mr. Gardner has contrived a small submarine whose movements are controlled by wireless telegraphy from a fixed point, such as the deck of a battleship. When the key of the transmitter is set in action, the ether waves are arrested by a receiver upon the weapon, and conveyed to a small electric motor, which is thus set in motion. It must be explained, however, that the energy for propelling the motor is not transmitted through the air, but the etheric waves control the action of the energy upon the little craft. The motor in turn drives a centrifugal governor. As the speed of the governors is increased, the force so generated is communicated to a series of switches, each of which represents an action to be controlled. By means of a chart the operator is guided in his manipulations of the transmitting key, in order to deviate his weapon from the straight course either to the right or left. Directly the pursuing boat comes within sufficiently close range, the operator opens a switch, and the 200 pounds of guncotton which the small crewless submarine carries is detonated."

OCTOBER 1898

"Mr. Charles F. Brush read a very important paper before the American Association for the Advancement of Science, in which he describes extracting from the atmosphere a gas which is lighter than hydrogen. The new substance has been called 'etherion.' Mr. Brush says that the ability of etherion to conduct heat is fully a hundred times as great as that of hydrogen. He also considers that the gas reaches out indefinitely into space."

OCTOBER 1902

"Dr. H. W. Wiley, Chief of the Division of Chemistry of the Department of Agriculture, will open in the autumn, under the authority of Congress, a kind of laboratory boarding house for the purpose of testing the effect of various preservatives, coloring matters, and food admixtures upon normal, healthy persons. The young men in the scientific bureaus of the Department of Agriculture will be drawn upon first, and after them the resident college students of the city of Washington. Dr. Wiley intends to ascertain the relative harmfulness of various substances as a part of the movement toward pure food legislation. The effect of borax on foods has not been quite definitely determined. The German government contends that the small amount of boric acid used in curing meat is not harmful. Dr. Wiley's experiments will either substantiate or refute that belief. Each boarder is to keep a diary and record of all facts concerning himself. He is to eat only what is set before him, and, in accordance with Scriptural injunction, is to ask no questions, for the sake of his conscience if not of his stomach."

DECEMBER 1902

"A device which does not seem to receive from its makers the attention which it merits is the sprag, the iron rod suspended from the rear axle to hold a car on a grade in case brakes do not operate or are not in use. Too often the sprags fitted to heavy large cars are altogether too slender for the purpose; often they are stout enough, but so short that the car would be certain to ride over them. It is not often that the sprag is needed, but when it is wanted the need is great and immediate, and not only the car, but the lives of its occupants may depend upon the apparently insignificant device."

SEPTEMBER 1903

"*The Anglo-Indian Review* summarizes an interesting account of the possible future applications of radium. In its industrial application we are somewhat restricted by the extremely limit supply of radium available, but it stated that a small fraction of an ounce properly employed would probably provide a good light sufficient for several rooms and would not require renewal during the present century. It has been calculated that the energy stored up in 1 gram of radium is sufficient to raise 500 tons weight a mile high. An ounce would, therefore, suffice to drive a 50-horse-power motor car at the rate of 30 miles an hour round the world."

FEBRUARY 1904

"The Technology Club of New York City recently held a radium banquet, in which the health of the Massachusetts Institute of Technology, whose alumni compose the association, was drunk in 'liquid sunshine.' A tiny tube of radium had been placed in water in a tiny cocktail glass. A magnesium wire was burned in a corner of the darkened room and in each glass there glowed a brilliant blue fluorescence. The toast to alma mater was then drunk standing."

SEPTEMBER 1904

"Many new thrills and novel sensations are being experienced by the guests at the St. Louis Exposition, and a company has undertaken to put up the great American refreshment, ice cream in the most novel and convenient form which has ever been devised. This company utilizes the collapsible tube in which paint has been so long sold for the use of artists. This tube was used first for this purpose and later came into favor for tooth paste, some forms of soap and similar commodities. The inventor is of the opinion that this invention will appeal to the great majority of visitors to the fair, for the reason that it will be a time saver."

DECEMBER 1904

"The alcohol habit has taken such a hold on the Russians that recently the Imperial Minister of Finance offered a prize of 50,000 rubles, which is equal to $25,750, for the discovery of some means by which the alcohol would be rendered so distasteful that it could not be consumed in this manner."

APRIL 1905

"Telephone tea parties are now in vogue on farm lines. There are telephone evening musicales. The accomplished contribute the programme, while others, scattered over an area of many miles, form the audience. The result is more satisfactory than the phonograph. A news service is one of the innovations of rural telephones. At seven o'clock in the evening a general call is rung. When each subscriber is at his instrument, the exact time is given; for instance: 'It is now one minute and a half past seven.' Then the weather conditions; then the late afternoon national and international news. McKinley's death was known to farmers ten miles from market towns as soon as in the cities. It is not too much to say that the telephone is working a revolution in rural life which in time will form an important chapter in sociology."

FEBRUARY 1909

"As a result of a lecture delivered by Sir Frederick Treves, the eminent British surgeon, in which he illustrated some practical curative results attained by the use of radium, a British Radium Institute has been founded for carrying out research operations in connection with the application of radium to surgery. In the course of his lecture Sir Frederick stated that radium can cure every form of naevus; will eradicate the terrible port-wine stain, which is probably one of the greatest disfigurements with which one can be afflicted; and will rid the patient of the pigmented mole and hairy mole."

MARCH 1909

"Prof. Muensterberg's 'machine for detecting lies,' technically known as a galvanic psychometer, has attracted much attention, despite the caustic comments with which it has been received in scientific circles. Many interesting results have been obtained with it by Dr. Veraguth of Zürich. A noise, a light, a touch, reading of an exciting novel, mental calculation or the recollection of some exciting incident, all produced—at the end of a few seconds, which may be called the latent period—a marked increase of the current."

NOVEMBER 1911

"Many of the most serious automobile accidents are due to a misunderstanding or ignorance of the driver's intentions and so a system of hand signaling has come into vogue, which, although crude, answers many purposes so long as the driver of one car is enabled to see that of another. At night such communication between the car operators is impossible. A rear signal has been devised for the purpose. It is electrically operated and consists of three lamps and a horn. A red lamp is lighted permanently, while above is a green lamp which being flashed, signifies that the driver is about to stop. To the right and left are white lamps, signifying his intention to turn to the right or left. These signals are all electric and are operated by buttons, conveniently placed for operation by the chauffeur. As any one of these signals is made, the horn is sounded to attract the attention of whoever may be following."

DECEMBER 1911

"Dr. Percival Lowell, of the Lowell Observatory at Flagstaff, Ariz., has been finally successful in obtaining photographs of Mars that establish beyond a doubt the reality of the canals of Mars. Heretofore, the value of the photos obtained by him in 1907 was questioned on account of their minuteness, being compared in size to the head of an ordinary pin. While the canals are plainly visible on the photographic plate, they will not bear printing processes. On examining the enlarged image through a screen by a stereopticon, an immense amount of elaborate detail appeared, several of the canals being plainly in evidence. As these photographs were taken a month or so before Mars had reached its nearest to our planet, we may look forward with interest to those which will be obtained on the date of nearest approach."

MARCH 1918

"Several European observers of the total lunar eclipse of July 4–5, 1917, have reported that the brightness of the lunar disk appeared much greater around the limb than near the center. These observations lead M. A. Nodon of Bordeaux to revive a suggestion that has sometimes been made to account for the brilliancy of certain lunar craters; viz., that the surface of the moon may possess a luminosity of its own in the nature of phosphorescence. In that case, perspective would increase the apparent luminosity toward the limb."

JUNE 1918

"The American concrete steamship *Faith,* the first large ocean-going vessel of concrete in the history of the world to make a successful voyage with cargo, completed her maiden and trial trip on June 2nd by tying up to the dock at the Canadian port of Vancouver. The *Faith,* loaded with a rough cargo of salt and ore, left San Francisco on May 22nd, touching at Tacoma and Seattle to unload. The voyage has been watched with anxiety and hope by men high in shipping circles of the Allies, for with the success of concrete ships proved, the way lay open to solving the ship shortage problem, as concrete ships can be built much more quickly and easily than either those of wood or steel."

JULY 1921

"It is rather startling when men of unquestioned scientific standing tell us that all of the tissues of the body are essentially immortal and that, barring accidents, we ought never to die. This is the newest evidence the science of medicine has to offer, and it is evidence, mind you, not theory. A skillful surgeon has been able to keep alive by artificial means, outside the animal, a bit of tissue for a longer time than the natural span of life of the animal itself. The remarkable thing is that the tissue is no longer subject to the influence of time and there is no doubt that if properly cared for it will live on indefinitely. The surgeon is Dr. Alexis Carrel of the Rockefeller Institute in New York and his experiment is with the heart of an embryo chick, which he has kept alive for more than eight years."

AUGUST 1921

"A French engineer, H. M. Melot, has put out an invention that from its form he calls the propelling trumpet. Capable of application to all sorts of vehicles, it is designed primarily for the airplane. The apparatus consists of a number of tubes ending with trumpet-like flares or nozzles. These are arranged in series in connection with a combustion chamber, where an explosive mixture of air and fuel is ignited as in ordinary engine practice. The exhaust gases of the combustion chamber are discharged into the series of nozzles. Both the pipe that effects this operation and the nozzles themselves are carefully designed to cause the expansion of the gases to occur under the best circumstances. It is the velocity of the exhaust and the velocity of expansion that, through reaction against the external air, drives the machine forward. The exhaust gases are discharged at a velocity of from 1,200 to 1,500 yards per second; at the entrance to each nozzle a certain amount of the outside air is drawn in and surrounds the jet of exhaust gas as perfectly as possible. The gas therefore gives up a part of its velocity to the air and causes a powerful suction action at the entrance to each nozzle."

SEPTEMBER 1922

"Professor Karl von Frisch has recently published in the *Münchener Medizinische Wochen-schrift* some observations on the means of communication employed by bees. He placed a dish of sugar solution on a table by an open window. Shortly after a chance bee had noted this and flown off with booty therefrom the dish was crowded with bees. By touching the back of each bee with a spot of color the experimenter then perceived that subsequent bees had been sent and not escorted. The conduct of the discoverer bee on its return to the hive was then noted. The bee first gave over its plunder to the other workers and then executed a curious dance, describing circles and other figures. Its audience watched attentively and attempted to touch it. When one of the marked bees succeeded in doing so, it at once made its exit and flew to the feeding place. It appears that there is some means of communication based on touch rather than on sight or hearing."

DECEMBER 1923

"Radio-acoustic marine direction-finding is a means of locating the position of a ship at sea by the emission of a radio 'dash' simultaneously with the firing of a small charge in the sea. A station on shore records the arrival of the radio signal, and also of the explosion wave at a number of hydrophones suitably disposed in known positions on the sea bed. The time of travel of the explosion wave, and hence the distance from the charge to each hydrophone, are indicated by a photographic recorder. It is possible to give a ship her location within a radius of half a mile, inside 10 minutes from receiving her request for a position. A nine-ounce charge can be located at a distance of 40 miles."

DECEMBER 1927

"At an address to the American Chemical Society at State College, Pa., Miss Elizabeth Wagner was presented as a model of a modern bride clad from head to toe, except for the soles of her slippers, in synthetic materials. Her dress was made of rayon fibers trimmed with rayon lace. The sleeves were of cellulose acetate fibers. Her tulle bridal veil was a nitrocellulose product. The orange blossoms were precipitated calcium carbonate coated with paraffin. Her stockings were of rayon. Her slippers were of rayon and metal threads, the metal a tin-copper alloy. Her beads were made of collodion with fish-scale essence as the iridescent material. Her prayer book had a celluloid back and the paper and ink were both chemical products. Even the traditional garter, embodying 'something old, something borrowed and something blue,' was made of rubber rendered adaptable by chemical treatment, covered with rayon and ornamented with rayon and metal roses. And as for the remainder of the bride's trousseau, she now has an unlimited number of synthetic textile fabrics from which to choose. Brocaded rayon velvets, rayon fabrics for sport wear, charming color effects made by combining two kinds of rayon with different dyeing properties and dresses and shawls embroidered with lustrous rayon threads, hats of rayon, of rayon plush or of cellophane, beads of collodion, of glass, of casein or of Bakelite—all these and more are to be had."

FEBRUARY 1931

"Again this year there is a strong trend toward eight-cylinder cars to replace sixes. Buick has entirely abandoned the six-cylinder engine in favor of the eight. Marmon is producing a 16-cylinder car for the first time and Cadillac a 12 to supplement its eights and its line of 16's announced last year. With the increased number of cylinders has come increased horsepower. Piston displacements of the larger Packard and of the Chrysler Imperial closely approach 400 cubic inches. In line with the increases in horse-power and displacement many cars have longer wheel bases. The Cadillac V-16 leads in this respect with a wheel base of 148 inches."

AUGUST 1935

"You can now acquire a full-size, life-long home simply by ordering from a catalog and telling the dealer where to erect it. When you take possession a month later, the refrigerator will be making ice cubes, the heating plant will be throwing cool, conditioned air through the rooms and as likely as not, the radio will be playing. This is the pre-fabricated house."

JUNE 1936

"The new German Zeppelin, *LZ-129*, has been christened the *Hindenburg* and has passed its flying tests with perfect success. If the wonderful record of its sister ship the *Graf Zeppelin* over the South Atlantic is borne in mind, there is little doubt that the same regularity and safety may be expected in the service of the new airship over the North Atlantic."

NOVEMBER 1936

"R. A. Lyttleton has advanced one of the perplexing problems of cosmogeny by suggesting a way of origin for the planets that avoids the hopeless dynamical difficulties that obsess the older theories. If the sun had originally another star revolving around it—perhaps at about the present distance of Uranus or Neptune—and a third star, passing by in space, collided with the companion, or at least passed exceedingly close to it, it is quite possible that both stars might fly away into space in different directions, leaving masses of ejected matter under the control of the sun's gravitation to condense into the planets."

JUNE 1938

"New patrons of a bank at Suffern, New York, are puzzled these days by a curtain of pale bluish light that falls between them and the tellers, along the grilled windows. For some reason bank patrons try to get as close to tellers as possible, and bank employees suffer more than an average number of colds. Invisible ultraviolet radiation emitted by slender 30-inch tubes at the tops of the grilles protect tellers in this bank from infection."

NOVEMBER 1939

"Flying the air-ways in specially designed and equipped ships, 'sky mail clerks' are now able to make deliveries and to pick up mail without the necessity of landing. The ground equipment consists merely of two steel poles, 30 feet high, set 60 feet apart. Each pole is topped by a brilliant orange marker. Stretched between the poles, and attached by spring clips, is a transfer rope from which is suspended the mail bag to be picked up. With a relatively simple system of grapple hooks, cables and winches, it is possible to deliver and pick up mail while flying at between 90 and 110 miles an hour."

OCTOBER 1940

"A powerful blast of air racing six times faster than a tropical hurricane recently blew out an electric arc powerful enough to light for an instant all the lamps in Chicago. Only a slight puff of smoke trailed off as a reminder of the once powerful arc. The new circuit breaker quenches in a few inches of space an arc that theoretically would have to be pulled out some 40 feet to be extinguished in ordinary air."

JANUARY 1945

"Perhaps the most interesting and promising of the proposed uses of glass-reinforced plastics are to be found in models for space-saving, structure-supporting, prefabricated kitchen and bathroom units. The two-sided assemblies, complete with full storage facilities, are intended to occupy a space only seven feet square, yet they are capable of supporting the entire structure of a house."

JANUARY 1947

"When the city-fathers of a municipality decide to spend some of the taxpayers' money for a new sewage-disposal or water-supply system, one type of piping, made from asbestos fibers and cement, is at or near the top of the list. It is free from various types of corrosion, and its internal smoothness keeps flow capacity at a peak through the years."

APRIL 1949

"A major offensive against malaria is to be launched by the World Health Organization, International Children's Emergency Fund, and the Food and Agriculture Organization. DDT has now made it possible to control the disease. WHO teams have been in Greece for a year, battling malaria with DDT and synthetic anti-malarials. Demonstration units have just arrived in Indo-China and Siam. Similar teams will be sent to Burma, Ceylon, India, Indonesia, Malaya, Pakistan and Yugoslavia. In southern Greece, three years of DDT treatment to eradicate malaria-bearing mosquitoes have reduced the malaria incidence from one million to 50,000 a year at an annual cost of 30 cents per person."

CHAPTER NINE

JUNE 1894

"The *Medical Record* tells of a woman in Ohio who utilized the high temperature of her phthisical husband for eight weeks before his death, by using him as an incubator for hens' eggs. She took 50 eggs, and wrapping each one in cotton batting, laid them alongside the body of her husband in the bed, he being unable to resist or move a limb. After three weeks she was rewarded with forty-six lively young chickens."

SEPTEMBER 1894

"The French War Office seems to be the target for all inventors, intelligent and otherwise. One invention takes the form of a captive shell, made to explode over fortresses, etc., and containing a small camera attached to a parachute. The enemy's fortifications would be photographed instantaneously, the apparatus hauled down like a kite, and the only remaining operation would be to develop the plates. Another inventor thinks that explosive bullets filled with pepper would have the two fold result of blinding the enemy and fostering French trade with its colonies."

FEBRUARY 1895

"At a place on the Mianus River, near Bedford, N.Y., known locally as the 'ten foot hole,' the stream widens out into a pool forty or fifty feet wide. In this pool there has formed a cake of ice about twenty-five feet in diameter and perfectly circular in shape. This cake is slowly revolving and is surrounded for about two-thirds of its circumference by stationary ice. There is a space of about three inches between the revolving cake and the stationary ice, except at the up stream side of the cake, where the water is open and the current quite swift. Each revolution takes about six minutes."

AUGUST 1895

"Niagara Falls will probably be the location of a factory for turning out electric men; not mesmerists or svengalis, but automatons that will run by electricity. They have built one up at a plant in Tonawanda; the man clothed in Continental uniform drags a heavy cart about the streets with some ease. Future models of electric men will be run by storage batteries and have a phonograph. The phonograph can expound the virtues of patent medicine or be used for political campaigns."

APRIL 1897

"It is said that 95 per cent of visual hallucinations in delirium tremens consist of snakes or worms. Investigation in the alcoholic wards of Bellevue Hospital with the ophthalmoscope reveals some interesting facts. In all sixteen cases examined the blood vessels of the retina were found to be dark—almost black—with congested blood. These blood vessels, which are so small and semitransparent in health, assume such a prominence that they are projected into the field of vision, and their movements seem like the twisting of snakes."

MAY 1897

"The visible sign of cobwebs and dust on a bottle of wine used to be taken as convincing evidence of age. Unfortunately, the Division of Entomology of the U.S. Department of Agriculture says that an industry has recently sprung up which consists of farming spiders for the purpose of stocking wine cellars, and thus securing a coating of cobwebs to new wine bottles, giving them the appearance of great age."

NOVEMBER 1897

"'Alpine misadventure' is a wide word, and includes victims whose sudden fall into a crevasse or mountain torrent is set down to 'loss of balance,' 'misplaced footing,' or one of many mishaps besetting the mountaineer, when syncope—fainting—due to cardiac lesion was the real cause. The hypothesis is strengthened by the death of a burgomeister of a Westphalian town, on the Furka Pass on the Rhone Glacier. The burgomeister, rising in his carriage to get a better view, had barely uttered, 'Oh! C'est magnifique!' when he dropped down dead. The altitude, the rarefied air, the tension—conditions inseparable from Alpine ascents—were too much for a 'chronic sufferer from weak heart.'"

JANUARY 1898

"The catalog of brilliant achievements of surgery must now include the operation performed by Dr. Carl Schlatter, of the University of Zurich, who has succeeded in extirpating the stomach of a woman. The patient is in good physical condition, having survived the operation three months. Anna Landis was a Swiss silk weaver, fifty-six years of age. She had abdominal pains, and on examination it was found that she had a large tumor, the whole stomach being hopelessly diseased. Dr. Schlatter conceived the daring and brilliant idea of removing the stomach and uniting the intestine with the oesophagus, forming a direct channel from the throat down through the intestines. The abdominal wound has healed rapidly and the women's appetite is now good, but she does not eat much at a time."

APRIL 1898

"According to a French writer named Petrie, twenty per cent of all cannibals eat the dead in order to glorify them; nineteen percent eat great warriors in order that they may inherit their courage, and eat dead children in order to renew their youth; ten per cent partake of their near relatives from religious motives, either in connection with initiatory rites or to glorify deities, and five per cent feast for hatred in order to avenge themselves upon their enemies. Those who devour human flesh because of famine are reckoned as eighteen per cent. In short, deducting all these, there remains only twenty-eight per cent who partake of human flesh because they prefer it to other means of alimentation."

MAY 1899

The "Amphibie" is the name M. Theodorides has christened his new nautico-terrestrial tricycle, which has recently been tried in France.

The tricycle is constructed entirely of aluminum, with the exception of the chain and certain other parts which require the use of steel. The wheels have enormous inflated rubber tires, which give them a diameter of 3.83 feet, and which make each wheel a water-tight float, buoying up the machine on the water.

The tricycle can be used indiscriminately on land or water, and although it does not run very rapidly, it may be of considerable use in special cases.

It weighs but 66 pounds and sinks, when fully loaded, to a depth of only 12 inches.

JUNE 1902

"A very novel idea has just been produced by Mr. Clarence M. Stiner, of New York City. He has designed a hairbrush for use in public places which at any time may be operated at a nominal expense to present a fresh, clean set of bristles for the user. The bristles are radially attached to hubs forming wheels, and the wheels are connected by a gearing. On the handle portion of the hairbrush is a mechanism for rotating the brush wheels. This mechanism can be started only on the insertion of a coin."

NOVEMBER 1902

"Professor Charles Frederick Holder, reporting on the abilities of various fish to make vocal sounds, writes: 'One of the most remarkable sound producers I have ever heard is a Haemulon in the Gulf of Mexico. When I took one of these fishes from the water it began to grunt:

"Oink-oink-oink"; now with one prolonged "o-i-n-k," then strung along rapidly, as though to intensify its agony; all the while it rolled its large eyes at me in a comical manner. No one in listening to such a remarkable outcry from a fish could refrain from wondering whether it had any significance: in other words, the impression was created that it was barely possible that the sounds were repeated in the water, and that they represented a very primitive attempt at vocal communication among fishes."

FEBRUARY 1912

"In a recent issue we published a description of an aviator's garment, so designed that it would belly out and constitute a parachute after a fall of a few feet. So convinced was the inventor, Franz Reichelt, an Austrian resident of Paris, of the merits of his garment that he determined to test it by jumping from the Eiffel Tower. On February 5 he donned his curious garment and proceeded to carry out his experiment, against the advice of M. Hervieu, an expert and the inventor of a similar device. He leaped from the first platform of the tower. The parachute failed to work and Reichelt was instantly killed by the fall of 180 feet. Reichelt's garment was provided with an enormous hood of silk gathered together on the back so that it resembled a knapsack. The hood was to be released by the wearer of the garment himself. Reichelt had made experiments with his garment in the courtyard of his house. A manikin had been used. The preliminary trials were for the most part unsatisfactory, according to accounts that we have received from Paris. It seems incredible that any man should venture on such a hazardous attempt and repeat it on so large a scale after failure."

JUNE 1916

"At a recent meeting of the Royal Institution in London, Sir James Dewar exhibited a remarkable soap bubble that he had blown a month before and which was still as perfect as when formed. It is described as a glowing sphere of iridescent color, showing no signs of 'blackness,' which is the prelude of collapse. The longevity of the bubble is described by Lord Rayleigh as a case of suspended gravitation, which is due to the fact that it was blown in and with clean air, free from motes, which appear to be the seeds of decay."

FEBRUARY 1919

"On January 19th last, Jules Vedrines, the famous French airman, set out from the aviation field at Issy-les-Moulineaux, notwithstanding a thick fog, and flew toward Paris. He flew rather low over the boulevards in order to get his bearings. On approaching the Galeries Lafayette, a large department store near the St. Lazare station, Vedrines shut off his engine and volplaned toward the roof. Skimming the parapet by a few inches, he made a spectacular landing, although the machine was slightly damaged. Vedrines won a prize of 25,000 francs ($5,000) for being the first airman to land on the roof of a house."

MAY 1938

"How far can one read a newspaper by the world's most powerful light? A group of technicians sought the answer to this question recently, when the 2,000,000,000-candlepower beacon atop the Colgate-Palmolive-Peet Building in Chicago was turned into the world's largest reading lamp for 90 minutes. Flying at 7,000 feet over Chicago, passengers on board a special United Air Lines *Mainliner* were able to read a newspaper by the light of the huge airway beacon at a distance of 27 miles."

FEBRUARY 1939

"Something unusual in aluminum safety equipment is the Christiansen safety valve for cows. A rather common complaint of cows is too great indulgence in alfalfa, with the result that the cow becomes badly bloated and in danger of death. The aluminum safety valve, with glass check ball, is permanently inserted in the cow's side. A cow that boasts this latest piece of spare equipment can eat its way through a field of alfalfa and, except for a slight whistle of gas through the valve, suffer no ill effects."

JUNE 1894

"Edward L. Still of New York City has devised an article of furniture which may be used as a lounge, right or left as desired, and which may also be employed as a bath tub or a wash tub, while it may likewise be made serviceable as a washstand or as a bed. The construction is exceedingly simple, and it may be easily and conveniently changed into any one of its several forms of use."

MAY 1894

One of the most entertaining as well as hygienic amusements is bowling. The exertion required to project the balls involves nearly all of the muscular system of the thorax. The arms, lungs, heart, back, and loins all respond to the movement, and the play is at once healthful and invigorating. For young people of both sexes it is particularly beneficial. It develops the limbs and chest, and imparts grace and flexibility to the body. But the practice of bowling is at present very limited, owing not to the cost of the appurtenances, but chiefly to the great length of the floor space required. A first-class single bowling alley costs $250, and requires a flooring 85 feet long and 6 feet wide. The practice of bowling at home in ordinary dwellings is, therefore, out of the question. Special houses for bowling are required, except when the cellars or basements of large buildings such as clubs or hotels, are made available.

The object of the present design is to modify the longitudinal dimensions of the bowling alley and adapt it, if possible, to the requirements of domestic life, in short, to make a bowling alley that may be used in the play room or other apartment of almost any good sized dwelling house. Instead of the long straight floor, a circular cycloidal pathway for the balls is provided, the track being thus, as it were, bunched up in the air, instead of being extended out in a straight line as a floor.

The balls are kept within the spiral pathway by centrifugal force, the principle of operation being the same as the well known spiral railway, in which the car sticks to the track, and the passengers keep their seats, although the car flies along bottom upward.

MARCH 1853

"Experiments have been lately made at Chicago to ascertain the amount of oxygen necessary to support life. Six hundred persons having been placed in a hall in one of the hotels of that city, all the doors and windows were closed. At the end of the third half-hour it was found unsafe to continue the experiment any longer."

FEBRUARY 1851

"Our moustached friends will be glad to learn that the wearing of moustaches is conducive to health. It affirms that the moustaches, acting as a part of the breathing apparatus, absorb the cold of the air before it enters the nostrils, and are consequently a preservative against consumption."

JUNE 1853

"The stone masons in Glasgow, Scotland, acting on the advice of Dr. Allison, of Edinburgh, have commenced wearing mustachios as a preservative against the injury done to the system by fine particles of sand, while they are engaged dressing stones. Custom may be against such natural preventatives: but if it is found that they are at all beneficial, we deem it the duty of some of our medical readers to recommend their adoption by millers, bakers and others similarly exposed."

AUGUST 1853

"That eminent chemist Justus Liebig says in his *Letters on Chemistry*, 'The quantity of soap consumed by a nation would be no inaccurate measure whereby to estimate its wealth and civilization.' By this measure we could justly claim for the United States the title of the wealthiest and most civilized nation in the world. Pillars of soap, busts of soap, windows of soap, soap of all colors, in all shapes, in all sizes and of all smells mark the vast extent of our soap manufacture. We are no doubt the best washed people in all creation."

DECEMBER 1853

"Mr. Goddard has arrived at the acme of aeronautic achievement in Paris. He has come down from a balloon in a parachute on horseback! Two years ago to *go up* in a balloon on horseback was a marvel. The parachute was immense, and the cords, extending from its edges to the framework that sustained the horse, were a hundred feet long. The umbrella was by some contrivance opened before the cord of connection with the balloon was cut in order to avoid, under the peculiar circumstances of the descent, the rapid fall that ensues till the silk unfurls. The aeronaut above (his brother) let him off at the height of a mile; the descent was easy and gentle."

MARCH 1855

"In some sections of the State of Mississippi the people, it is reported, have pulled down a number of miles of telegraph wires, because some learned ignoramus had demonstrated to them that the long drought in these regions was caused by these wires carrying off the lightning, which used to bring rains."

JUNE 1855

"The Academy of Sciences in Paris have been investigating the causes which almost invariably make the west end of a city grow more and become more fashionable than the east. 'It arises from the atmospheric pressure,' answers the Academy of Sciences. The wind which causes the greatest ascension of the barometric column is that of the east, and that which lowers it most is the west. When the latter blows, it has the convenience of carrying with it to the eastern parts of a town all the deleterious gases which it meets in its passage over the western parts; and the inhabitants of the eastern part of a town have to support not only their own smoke and miasma, but those of the western part of the town, brought to them by the west winds. When, on the contrary, the east wind blows, it purifies the air by causing the pernicious emanations, which it cannot drive to the west, to ascend. The deduction from this law is, that the western part of a city is the best place of residence for persons of delicate health."

JUNE 1857

"The camels which were imported by our Government from Arabia are reported to be doing well in Texas, and as likely to become acclimated as horses. Several native American camels have been born, and others are expected. The only question relates to the quality of the young animals. It is said that the Turks look with suspicion on our efforts to contract for building railroads in their country, while we are at the same time buying their camels. They say we want to get rid of our railroads and adopt their 'improvement.'"

SEPTEMBER 1858

"We see it stated that the medical attendant of the Princess of Gothland asserts that hoopskirts are the cause of accouchements lately becoming so dangerous and difficult. He adds that this fashion is the source of a vast number of chills, the consequences of which are, in many cases, mortal. We have always thought that hooped skirts of reasonable bounds were not only an adornment to the persons of the fair wearers, but, on account of their ventilating character, actually beneficial to health. Ladies generally evidently think so, and as they are the actual sovereigns of creation and will wear what suits them, we doubt whether this statement will have any effect."

NOVEMBER 1858

"About two years since New York City was visited by Alexis St. Martin, of Canada, who has an opening in his abdomen (the result of a gunshot wound) through which his stomach can be examined, and the operations of digestion observed. His case has hitherto been considered the most wonderful in the world, but one more wonderful than that of St. Martin is now in New York. During the past week, M. Groax, a native of Hamburg, exhibited to the faculty in the Columbia University Medical College his own beating heart, in the same manner that St. Martin did his stomach. This case however is a natural phenomenon, Mr. Groax having been born with a slit in his breast, by which his heart and a part of his lungs can be observed."

JUNE 1877

"A new mania is at hand, to wit the celery cure. 'Celery is the greatest food in the world for the nerves,' says one of our contemporaries, and the information is traveling the length and breadth of the land. It is fashionable nowadays to call every ailment that flesh is heir to a nervous disease, and where our ancestors would have resorted to such homely remedies as a hot drink and simple cathartics, the present practice demands chloral, bromides, quinine, strychnine and phosphates. Of course, celery is pleasanter to take than most drugs, and now that it is brought forward as a new nervine plenty of people will use it. As it can do no harm, and indeed may actually work good by checking the too prevalent consumption of 'nervous specifics,' the mania is rather a benefit than otherwise and should doubtless be encouraged."

FEBRUARY 1848

"In Franconia, N.H., the weather is said to be so cold that the natives lather their faces and run out of doors, where the wind cuts their beards off."

AUGUST 1897

"An inhabitant of the Scilly Islands was struck by the fact that the rats there seemed to prosper greatly, although the place is very barren. He resolved to investigate the cause of this, and digging up some of the nests by the seashore, found that the rats had dragged crabs into their holes, and, in order to prevent their escape, had bitten off their legs. No doubt the prey had been seized at low tide and brought home."

JUNE 1846

"Among the fancy inventions recently introduced is a genteel bee-hive for the parlor, invented by Mr. J. A. Cutting, of Boston. It is finished in the style of elegant cabinet furniture, and about the size of a bureau, with glass doors in front, through which the operations of the 'busy bee' can be observed. Meanwhile, the bees, not intimidated by contiguity with equally civil though less industrious society, being furnished with a private entrance through the walls of the house, pursue their avocation with security."

JULY 1848

"We were escorted through a crowd of wondering Dyaks, to a house in the centre of the village. The structure was round and well ventilated by port-holes in the pointed roof. We ascended to the room above and were taken a-back at finding that we were in the head house, as it is called, and that the beams were lined with human heads, all hanging by a small line passed through the top of the skull. They were painted in the most fantastic and hideous manner. However, the first impression occasioned by this very unusual sight soon wore off, and we succeeded in making an excellent dinner, in company with these gentlemen.—Frank Marryat [*Excerpted from Marryat's* Borneo and the Indian Archipelago, *published in London in 1848.*]

NOVEMBER 1848

"The Woodstock, Vt., Mercury says: 'We gave some account a few weeks ago of the astonishing case of Mr. Gage, foreman of the railroad in Cavendish, who in preparing a charge for blasting a rock had an iron bar driven through his head, entering through his cheek and passing out at the top with a force that carried the bar some yards, after performing its wonderful journey through skull and brains. We refer to this case again to say that the patient not only survives but is much improved. He is likely to have no visible injury but the loss of an eye."

APRIL 1849

"From the Mount Hope Institute on the Insane, Dr. W. H. Stokes says, in respect to moral insanity: 'Another fertile source of this species of derangement appears to be an undue indulgence in the perusal of the numerous works of fiction, with which the press is so prolific of late years, and which are sown widely over the land, with the effect of vitiating the taste and corrupting the morals of the young. Parents cannot too cautiously guard their young daughters against this pernicious practice."

JULY 1849

"It is a fact well known to those who have visited the mountainous regions of Syria, Palestine, and the Peninsula of Sinai, that the camel is as serviceable on rough mountain paths as in the moving sand of the desert. The tough soles of the camel's feet are affected neither by the burning sand nor by sharp-edged stones. There is no reason why the camel should not be as serviceable to man on the Prairies of Texas and the mountain region of Mexico, New Mexico, and California, as in the corresponding tracts of the Old World."

AUGUST 1862

"We extract the following from Charles Dickens in the magazine that he edits, *All the Year Round*: 'We do not all come out of the photographic studio alike unhappy. There are those to whom the process does justice, as well as those to whom it does injustice. I have myself sat on two occasions for one of these portraits. On the first I was simply occupied in keeping still and presenting a tolerable favorable view; but the result was so tame and unimposing a picture that I determined on the next occasion to throw more intellect into the thing, and finding a certain richly gilded curtain tassel convenient to my gaze, I gave it a look of such piercing scrutiny, and so withered and blasted it with the energy of my regard, that I almost wonder it did not sink beneath the trial. That look has, I am happy to say, been reproduced faithfully, and no one could see the portrait without giving its original credit for immense penetration, energy and strength of character, and a keen and piercing wit.'"

APRIL 1866

"Rats are as plentiful in Paris as in London, and they are often the victims of physiological experiments. M. Bert, for example, gained the prize in experimental physiology for removing their tails from their natural position and grafting them upon all sorts of odd places—the middle of the back of the animal, for instance, and even in the cavity of the peritoneum. M. Bert made one very curious observation. He succeeded in uniting the small end of the tail to the body and found that the large extremity, which was free, recovered its sensibility, thus showing that the nerves will convey sensation in a direction inverse to that in which they act under normal circumstances."

DECEMBER 1867

"Some time ago the death of a young lady passenger, Miss Stainsby, in one of the cars of the London underground railway was reported caused, as then alleged, by suffocation due to the bad state of the air in the tunnels. A legal investigation ensued from which it now appears that one of the causes of her death was tight lacing. Prof. Rodgers, lecturer on medical jurisprudence and on chemistry, was the first witness, and at his request the evidence of Dr. Popham as to the appearance of the body was read to him. Dr. Popham added that he had found the deceased was tightly laced and that the result would be to compress her chest and impede the free action of her lungs. Prof. Rodgers said he had examined samples of air taken on four different occasions from the tunnels of the Metropolitan Railway, and also from various other tunnels. The slight deficiency of oxygen which he found would not act injuriously, even upon delicate persons, passing as they did rapidly through the tunnel in trains. He thought that under the circumstances under which the deceased had entered the train—that was to say, considering that she had eaten heartily, was tightly laced, had a diseased heart and was already faint before she entered the tunnel—her death had resulted from natural causes. The jury heard other evidence and then, without hesitation, brought in a verdict: 'Died from natural causes.'"

MARCH 1868

"Prof. De la Rive of Geneva has contrived an instrument for measuring the transparency of the atmosphere. The inventor agrees with Pasteur, who supposes that the light dry fog which under certain conditions of the air intercepts the light is caused by myriads of organic germs floating near the earth, which are washed to the earth by the heavy rains or are destroyed by severe frosts, thus accounting for the clearness of the atmosphere at these times."

AUGUST 1868

"The *Revue Populaire* of Paris gives an account of some very curious experiments made by Dr. Claude Bernard. If oxygenized blood be injected into the arteries of the neck immediately after decapitation, warmth and sensibility return, the eye gets animated and displays such perception that an object shaken before it will cause winking of the eyelids and movements of eyeballs as though to avoid injury."

SEPTEMBER 1868

"It is a commonly received notion that hard study is the unhealthy element of college life. But from tables of mortality of Harvard University collected by Professor Pierce from the last triennial catalogue, it is clearly demonstrated that the excess of deaths for the first 10 years after graduation is found in that portion of each class inferior in scholarship. Every one who has seen the curriculum knows that where Æschylus and political economy injures one, late hours and rum punches use up a dozen, and that the two little fingers are heavier that the loins of Euclid. Dissipation is a swift and sure destroyer, and every young man who follows it is, as the early flower, exposed to untimely frost. A few hours of sleep each night, high living and plenty of 'smashes' make war upon every function of the human body. The brain, the heart, the lungs, the liver, the spine, the limbs, the bones, the flesh, every part and faculty are overtasked, worn and weakened by the terrific energy of passion loosed from restraint until, like a dilapidated mansion, the 'earthly house of this tabernacle' fall into ruinous decay."

MARCH 1869

The single-wheel velocipede, has received a palpable body and a "local habitation and name" by the enterprise of the inventor of the machine herewith represented. Queer and odd as may be the appearance of the concern, Mr. Hemmings says that his son of thirteen years old propels one of these machines of five feet diameter at a pace that keeps up with good roadsters and does not allow them to pass him. The greyhound is not able to keep up with the rider of this novel velocipede, but his master is compelled to reverse his motion and throw the driving friction wheel back of the center of gravity.

The main wheel has a double rim, or has two concentric rims, the inner face of the inner one having a projecting lip for keeping the friction rollers and the friction driver in place, each of these being correspondingly grooved on their peripheries. The frame on which the rider sits sustains these friction wheels in double parallel arms, on the front one of which is mounted a double pulley, with belts passing to small pulleys on the axis of the driving wheel. This double wheel is driven, as seen, by cranks turned by the hands. The friction of the lower wheel on the surface of the inner rim of the main wheel is the immediate means of propulsion. A small binding wheel, seen between the rider's legs, serves to keep the bands or belts tight. The steering is effected either by inclining the body to one side or the other, or by the foot impinging on the ground, the stirrups being hung low for this purpose. By throwing the weight on these stirrups, the binding wheel may be brought more powerfully down on the belts. Over the rider's head is an awning, and there is also a shield in front of his body to keep the clothes from being soiled by mud and wet.

JULY 1870

"At a celebrated observatory in the suburbs of London a visitor was desirous of observing a celestial object which was nearly overhead, and having the run of the observatory at the moment, he directed the telescope towards the star, set the clockwork in motion and placed himself on his back in the observing frame attached to the floor. Intent on observing the star, our astronomer failed to notice that the movement of the eyepiece was gradually imprisoning him. His head was fixed by the headrest, and the eye-tube was beginning to press with more and more force against his eye. Fortunately his cries for assistance were heard and the clockwork was stopped. Otherwise he would have had as good reason to complain as the celebrated astronomer Struve in the case of the Pulkova refractor, which Struve said, was justly called a refractor since it had twice broken his leg."

AUGUST 1870

"Dr. A. Cabe of Lyons, France, writes that he had in his practice a very obstinate case of constipation in a female subject 80 years of age, who for 60 years had suffered in consequence of a severe attack of dysentery encountered in her youth. The patient having had no passage for 40 days, the doctor tried to induce a contraction of the intestines by the application of electricity. He inserted the negative pole of a Gaiffe battery into the rectum and applied the positive to the navel, and in the course of two minutes the results were completely satisfactory."

DECEMBER 1870

"Dr. H. C. Wood, Jr., has written an essay, which he read before the American Philosophical Society, in which he records some experiments with an article of hemp grown in Kentucky. He took an alcoholic extract made from the dried leaves, swallowing at a dose nearly all of the product of an ounce and a half of the leaves, with the effect of profound hemp intoxication. Other trials he has made with it convince him that it has more power than the hemp brought from India. The native plant will always be more reliable than the imported, from the certainty of freshness, while the cost of it is hardly anything."

SEPTEMBER 1872

"We regret to learn that the Allegheny Observatory at Pittsburgh has suffered a serious loss by the depredations of thieves, who recently broke into the dome room and carried off the object glass of the great equatorial telescope. This lens was the most desirable piece of property in the establishment, its value being $4,000. Nothing else was stolen; it is therefore evident that the robbers knew what they were about. The lens was one of rare excellence, 13 inches in diameter, and the third largest, we believe, in the United States. Its loss is keenly felt by Professor Samuel Pierpont Langley, as it of course renders the telescope useless."

DECEMBER 1872

"The object glass of the Allegheny Observatory at Pittsburgh, Pa., which was mysteriously stolen three months ago, has been found and restored to its place in good order. Its value was $4,000. It was carried off by an intelligent sort of thief who probably expected that a reward would be offered for its return, enabling him to make a little money by the operation. He deposited it in a stable, and his pal stole it from him. It finally came into the possession of the police and was carried home."

AUGUST 1884

"It is not generally known that there is an American town in the realms of the Czar, yet such is a fact, it being near Moreton Bay in Kamtschatka. The colony has been formed, gradually, by immigrants attracted by the establishment of important lumbering operations, including saw mills, by an American company, and the town itself has so far been practically ignored. It is not on any known map and does not appear in the Russian real-estate register or on any tax list. The consequence is that the inhabitants pay no kind of tax and, until recently at least, have remained independent of the Russian authorities."

NOVEMBER 1884

On the evening of October 31, this city was favored with one of the most unique and attractive displays ever seen in a torchlight procession—that necessary adjunct to a presidential campaign, which brings into active play the inventive genius of party managers and enthusiastic followers. That an electric lighting plant, complete in every detail, and in full operation, can be moved at the uneven pace of a procession over the rough paving of a street, without interrupting the current or in any degree changing the brilliancy and steadiness of the light, is a fact which, while of interest to the scientific world, clearly shows the perfection to which electric lighting machinery has been brought.

The work of preparing the display was done by the Edison Electric Lighting Company, the expense being defrayed by its own employes. Placed upon the forward part of a large truck was a dynamo—a 200 ampere machine—behind which was a 40 horse power engine of the New York Safety Steam Power Company; a belt led from the engine to a pulley on the armature shaft. Secured to the truck was the pole of one of the largest steam fire engines built by the Clapp and Jones Company. The electricians in charge of the display felt assured of the successful working of their dynamo and engine, and in order to have an ample supply of steam, they obtained the fire engine, which they knew to be a rapid and reliable generator while in motion. Extending from this boiler to the engine were two flexible pipes, one leading to the steam chest and the other carrying the exhaust. The latter pipe was provided with a three-way valve, by means of which the steam could be directed either into the smokestack to increase the draught, or into the open air. Following the fire engine were two ordinary watering tanks, holding 950 gallons, which were connected to the feed pump by lines of hose. Between the tanks were the coal carts. The dynamo truck was drawn by six of the Herring Safe Company's mammoth horses, arranged tandem and guided solely by the word of the driver.

Extending from a switch board on the floor of the truck were four covered copper wires, two of which led to a rope upon one side of the truck and the other two to a rope upon the other side. This rope was 1,200 feet long, and was extended up and down the procession so as to form a hollow square, in the center of which was the machinery, before and behind which marched bodies of men. Placed at each five feet on the rope was an ordinary cut out, or lamp receptacle, slightly changed to suit the requirements of this work, and within which screwed a safety catch carrying two wires, which led up the sleeve and through the back of the helmet to an incandescent lamp of 16 candle power. Wires also led to lamps hung upon the hames of each of the trucks. The leader of the procession, on horseback, carried a staff surmounted by a 200 candle power light. Altogether there were some 300 lamps distributed along the rope and upon the trucks.

Upon the first and last part of the line of march every part of the plant worked most admirably, and the illuminations was intense and beautiful, the light flooding every nook and cranny in the streets passed through. But in the intervening distance, which chanced to be lined with people who were particularly anxious to witness the electric light display, this portion of the parade was conspicuous solely on account of the darkness that prevaded it. This interruption was caused by mud from the water tanks clogging the hose leading to the pump. All went well after the hose had been cleaned.

JANUARY 1886

"Starvation, semi-starvation, 'banting,' alkalies, purgatives, Turkish baths, exercise and the thousand and one ways of reducing corpulency to respectable dimensions still leave a large section of our stout population in despair. M. Germain See comes to the rescue. 'Oh, ye massive fat ones desiring to be made lean, eat not much meat, but drink enormously of tea.'"

MARCH 1887

"At a meeting of the Caucasian Medical Society, Dr. A. P. Astvatzatur-off of Tiflis drew attention to the danger of infection arising from the promiscuous use of the mouthpieces of public telephones. To prevent any accident of the kind, he recommends that the mouthpiece should be disinfected every time after or, still better, before it is used. In other words, some disinfectant fluid should be kept at every telephone station and the speaker should dip the mouthpiece into it."

JULY 1888

"The women in the Sultan's seraglio, at Constantinople, have just been vaccinated, to the number of 150. The operation took place in a large hall, under the superintendence of four gigantic eunuchs. The Italian surgeon to whom the task was confided was stationed in front of a huge screen, and the women were concealed behind it. A hole had been made in the center of the screen, just large enough to allow an arm to pass through."

DECEMBER 1888

"C. W. Oldreive lately accomplished the task of walking on the water of the Hudson River from Albany to New York. Distance about 150 miles, wager $500. His average progress was twenty-four miles a day. He always went with the tide. The shoes he wore are of cedar lined with brass. They are five feet long and a foot wide."

JANUARY 1890

"The use of the Colorado Midland's rotary snow shovel on the Denver, Texas, and Fort Worth seems to have created a mild sensation. A local paper says: 'It was put to work in a big cut where the snow was about 20 feet deep. Around the center of the cut, a strange sight was witnessed. Those who were standing on either side of the plow were suddenly deluged with a shower of beef steaks. On all sides fell porterhouse, sirloin, round steaks, small steaks, shoulder steaks, with occasionally a slice of liver or a nicely cut rib roast. Investigation disclosed the fact that a herd of Texas cattle had crowded into the cut and had frozen and been buried in the drifts.'"

APRIL 1891

"A method of extracting teeth without pain was recently demonstrated in London. An electrical arrangement employs a couple of bichromate cells and a Ruhmkorff's coil to which is attached a commutator of extreme sensitiveness. A patient takes the handles of the battery in his hands. One of these is connected with the negative pole. The positive is divided into two, so that one of the divisions is connected with the handle and a wire from the other division is screwed into the handle of the tooth forceps. When the patient takes hold of the handles the current is gradually increased in intensity until the patient can bear no more; then, while the forceps are being introduced, the current is turned off for a second, and on again. 'Had you no pain?' asked our representative of the patient when the roots of the bicuspid were held up to view. 'Not a bit; I only felt the grip.'"

MAY 1891

"On the evening of April 25 last, during a violent thunder storm, the lightning struck the lightning rod until it came to a defective insulator, then entered the house, striking Mr. Roode about half an inch back of the ear and burning its way through the entire length of his body, then through a wool mattress, splitting a hard maple bed-stead, afterward passing through various parts of the house until it reached the water pipe. Mr. Roode regained consciousness and is on the road to recovery. His body is now so heavily charged with electricity that he can impart to any one an electric shock equal to that received from a powerful battery."

JUNE 1893

"An instance of rare presence of mind attended by success in the use of an antidote to poisoning occurred recently at Sag Harbor, N.Y.

Flora Sterling, the five-year old daughter of Dr. Sterling, while playing about the house found a bottle which had formerly contained citrate of magnesia and still bore the label. The child put it up to her lips and took a long swallow. With a scream she dropped the bottle and began to clutch her little throat in an agony of pain. Her father, who had heard her screams, found that what the little one had taken for citrate of magnesia was oxalic acid. Seeing that not a moment was to be lost, if he wished to save the child's life, the doctor looked about for an alkaline antidote. Seizing his penknife the doctor sprang to the whitewashed wall and scraped some of the lime into his hand. This he threw into the glass partly filled with water, and poured the mixture down the almost dying child's throat. The antidote took effect at once."